HARLEY-DAVIDSON *RACING* 1934–1986

Allan Girdler

ECHO POINT BOOKS & MEDIA, LLC

Copyright © 2014 Allan Girdler

Published by Echo Point Books & Media,
www.EchoPointBooks.com

ISBN: 978-1-62654-932-6

Cover design by Adrienne Nunez,
Echo Point Books & Media

Manufacturers and model names used in this text are for identification purposes only
and are the property of the trademark holder. This is not an official publication.

Printed in the U.S.A.

Contents

	Acknowledgments	5
	Preface	7
	Introduction	9
1	The birth of showroom stock (1934-51)	15
2	Reign of the flatheads (1952-69)	37
3	Lightweights (1961-86)	89
4	Modern times (1970-86)	120
	Index	189

Acknowledgments

Speaking just for myself, here's another cliche: Without the help of Brad and Len Andres, Carroll Resweber, Brent Thompson, Clyde Denzer, Bill Werner, Jerry Hatfield, Steve Wright, L. J. Jones, Bill Milburn, Buzz Buzzelli, Pieter Zylstra, Bob Della Santina, Steve Storz, Bill Hoecker, Tex Peel, Mert Lawwill, Mark Brelsford and the omnipotent Mr. O'Brien, this book wouldn't be here.

Preface

Carroll Resweber has gray hair, a game leg, a game arm, a grin as wide as his native Texas and a Texas talent for the spoken word. We sat in his kitchen till the sun went down, Resweber describing those early days: the best race he didn't win, the tricks played on the other riders, his life since his terrible crash in 1962. Today, Resweber is a machinist and welder. He earns a good living and works no harder than he has to, but he has nothing now he wouldn't have had anyway, even if he'd never been four-times national champion.

Except for the good times, that is. His eyes sparkle when he talks about racing. His face lights up, his hands dart back and forth as he runs hard into the first turn, banks the machine and gets back on the gas.

When it was time to leave I thanked Resweber not just for the facts and his time, but mostly for his enthusiasm.

We racing fans are fans because we can't do it. We lean on the fence and watch the impossible performed before our eyes. Then, in club races or in private we try it ourselves and learn, more or less painfully, that we were born to watch.

But Lord, how we wish we could do it!

So it means a lot to me to listen to a racer who appreciates his talent, who had fun doing well what I can't do at all. Nor, I suspect, am I the only one: Inside every reporter, tuner, team manager and photographer I know is a failed racer.

On behalf of the railbird majority, the millions of us who *can't*, this book is dedicated to the talented handful who *can*, and who in so doing let the rest of us forget, for twenty-five laps or so, that we'd rather be out there ourselves.

Introduction

William Harley and Arthur and Walter Davidson probably would read this book with mixed emotions. One hopes they'd begin with pride, in that Harley and the Davidsons were the founders of the company that bears their names. On the other hand, this is a racing history. Racing machines, factory teams and record speeds weren't what the founders had in mind when they began tinkering with engines at the turn of the century, or when they knocked together a shed behind the Davidson family home in Milwaukee and began building motorcycles in 1903.

But never mind that. To appreciate Harley-Davidson and racing today, and to understand how and why racers do what they do, we need to look at how it all began.

Tinkering is mostly how it began. At the dawn of the motorized age, neither trained engineers nor backyard visionaries really knew much about internal combustion, or how to harness this new power source to a wagon or a bicycle.

Speed wasn't the issue. As the racing maxim puts it, To finish first, first you must finish. Going fast had to wait until the pioneers learned how to make the things go at all, and then to keep them going.

So those first Harley-Davidsons had one cylinder. They were built as motorcycles, stronger if heavier than the motorized bicycles from rival pioneers. The engines were so efficiently muffled that the Harley was known as the "Silent Grey Fellow," and so stoutly constructed that in 1912 Harley ads pointed out that "Harley-Davidsons sold ten years ago are still running and giving satisfaction."

The buying public was at least as skeptical then as it is now. Jeering onlookers did in fact shout "Get a horse!" and pioneer motorists did get out and get under.

So it made sense for Walter Davidson to enter the 1908 Long Island Endurance Run, one of eighty-four riders representing twenty-two makes. He won, with a perfect score. Then came the economy contest, which he also won, with 188 mpg. In 1909 Davidson and three other Harley riders entered an endurance run from Cleveland to Indianapolis, which they finished with perfect scores and the team prize.

That should have proved the point, and it would have except that the point had been changed. Motorcycles (and cars) became reliable transportation more quickly than even the dreamers could have predicted. Motorcycle racing became a highly popular spectator sport, on closed tracks and with specialized equipment. It was thrilling stuff, dangerous and not closely related to what ordinary people (customers) did on the road.

So Harley-Davidson Motor Company took no official part in those early professional races. There was no team, no works machines, no hired riders. (Lest this sound preachyperfect, let the record show the factory was willing to advertise wins scored by racers whom the factory wasn't willing to support.)

And the first of motorcycle racing's several Golden Ages began.

Because it was easy to put a second cylinder on the crankcase of a single-cylinder engine, and because the vee fit so well into the bicycle-based frames of the day, the V-twin became the second popular type of engine used in those early days. There was even some standardization: The singles displaced 30.50 cu. in., or 500 cc, and the V-twins were 61 cu. in., or 1000 cc, displacements that have been common ever since.

Exotic motorcycles also appeared very early. Pioneer biker and aviator Glenn Curtiss, for instance, stuck yet another barrel on a V-twin's crankcase and came up with the first triple. He later turned a row of four V-twins sideways, tacked them together into a V-8, bolted the engine into a motorcycle and was clocked at 137 mph.

Harley team riders at speed on the Chicago board track, 1927. Harley-Davidson

That was in 1907. The world's race organizing bodies were based in Europe so naturally they refused to recognize the times. But that's another part of the story.

The point here is that every modern feature, be it electric starting, full suspension, multiple valves, cylinders, shaft drive, overhead cams, water cooling or two-strokes, had appeared on motorcycles before World War I.

So had the factory team. In Harley's case the team first appeared on July 4 (a lucky accident?), 1914, for the 300 mile race at Dodge City, Kansas. Six pocket valve (or ioe, intake over exhaust) twins were entered; two finished, both well off the pace.

Except for the 1915 Dodge City race, the team had eight-valve twins, real racing machines. The team, by then known as the Wrecking Crew, came in first, second, fourth, fifth, sixth and seventh. An Excelsior was third, while the Indian, Pope, Cyclone and Emblem teams were nowhere.

Harley-Davidson's position in racing was firm, solid, even overpowering. To understand this success we must look at several factors that will influence racing.

One was the machines. The racing Harleys were "full race," as people would come to say later. The eight-valve twins (and four-valve singles, effectively the same machine but with one cylinder removed and the flywheels rebalanced) bore faint resemblance to what Harley sold to the public.

Another factor was price. The 61 cu. in. twin had a retail price of $1,500 fob Milwaukee. The 30.50 single sold for $1,400.

Or, didn't sell. On purpose. Rival Indian offered its racer, the overhead valve Model H, which was also 61 twin or 30.50 single, four valves per cylinder, for $350 or $300, respectively.

One needn't do much head-scratching before realizing that Harley didn't sell many eight-valve racers because it didn't *want* to. Once into racing, the company went as heavily as possible, building bikes for, and only for, the factory team. They were nominally offered to the public as a gesture in recognition that the governing and sanctioning body encouraged the factories to encourage the sport, but had no way to enforce their encouragement. This had severe repercussions, as we'll see.

But first, the sanctioning bodies. This is complicated, as the details have been lost with time, but when motorcycles first appeared and ran against each other, there came into being a club, an organizing body called the Federation of American Motorcyclists. The FAM began putting on endurance runs in 1904 and progressed to sanctioning races. It became a truly national organization and was able to establish records and classes, such as using the 61s for mile tracks and road races, the 30.50s for half miles and short tracks and banked speedways. And the FAM recognized national champions, professional and amateur.

The FAM folded in 1919. So the sanctioning of races, recognition of championships and records, licensing of riders and so on was taken over by the Motorcycles and Allied Trades Association. The M&ATA was as the name implies, a trade group. The members were the motorcycle factories and the makers of various parts and accessories. They surely were intelligent and upright men, but just as surely, their interest in racing was professional, rather than sporting.

Professional racer Ray Garner, who was there, later said that a bunch of Chicago riders felt the companies should have one organization and the riders should have another. So six of the racers banded together and formed the American Motorcycle Association (AMA) in 1923.

This became the official sanctioning body. Instead of having one national championship meet and proclaiming the winners of that meet the national champions, the AMA granted a selection of national titles. One track was allowed to have the ten-mile national championship race. Another mile track would get the twenty-five-mile title, still another track had the five-mile race on the half mile and so forth. There was no national champion rider as such, although in time the Springfield, Illinois, mile became *the* mile race, and the winner of that race got to wear the AMA's No. 1 plate, while Daytona was *the* road race, and Peoria, Illinois, was *the* TT race. (We'll get back to the definitions of road race and TT shortly.)

The AMA, as originally founded, lasted until 1928, when the riders for whom it was formed failed to come through with their memberships. Because the sport and the business needed a club, a sanctioning body, Harley-Davidson, Indian and Schwinn put up the money to keep the AMA going.

This cooperative backing will become important. First, it's true that the motorcycle companies didn't save the club for noble, sporting reasons. But it's equally true that the factories didn't seize control of the racing body. In fact, they grumbled about having to bail out those who should have been willing to help themselves.

But, the big thing here was that the factories owned and controlled the club that sanctioned the races and wrote the rules. Further, for reasons lost in history, they did it at one remove and got red in the face and slammed down the phone when asked if they had the control. Poor public relations, we'd say now, except of course they didn't know the term back then, never mind being polite to the public.

About the events themselves: As often as not, road racing was done around a closed course, sometimes dirt and sometimes paved, but almost always a regular road borrowed for the occasion. If the course was permanent and irregular, that is with turns in both directions, track owners and promoters liked to include a hill, which grew into a jump. This was thought to resemble the horse racing done over jumps as well as around irregular courses. So, in the United States, races with turns in two directions and at least one jump became known as TT, for Tourist Trophy, and/or TT Steeplechase, like the cross-country horse races. In fact TT and road racing were much more like

Curley Fredericks on his ioe 61 cu. in. twin. Front forks are an early version on Harley's springer forks, leading link with perhaps an inch of travel. These bikes hit 120 mph on the banking, and they had no brakes and very brave riders, many of whom were killed or injured. Harley-Davidson

each other in America than either venue was like racing in other parts of the world.

Meanwhile, speeds became incredible and rules evolved to slow the races down, or try to. In 1921, a 61 cu. in. Harley ioe twin ran around a mile board track at 107.78 mph. By 1926 the record was 120.3 mph, on a 1.25 mi. banked board track, on a 61 twin with no brakes and vestigial front suspension and with tires as thick as your thumb.

Introducing a common theme here, in 1924 *American Motorcyclist* magazine editorialized that a four-valve 500 cc single-cylinder racer "can be handled properly by fewer than ten riders in the U.S.... roaring into the short corners of a half-mile track at 65-70 mph is too fast."

And sure enough, in 1926 a 21 cu. in., or 350 cc, class was created to slow down speeds on the half miles.

This brings up another point. There were several displacements, as in the 350 singles, the 500 singles, the 750 and 1000 cc twins, even the 1000 cc fours used in hillclimbs. There were limits on displacement, but that was nearly the only limit.

Also interesting was the AMA's affection for initials. The top racers, the favored few who earned places on the factory teams and thus could earn a living by racing, held Class A licenses. The guys on the next level, who rode for money but had to hold down jobs at the same time, were Class B riders.

Machines built for racing were Class A machines. This meant they were virtually without limit. The rules might say only 21 cu. in. or 45 cu. in., or whatever, but there was nothing in the book about number of cylinders or arrangement of the valve gear.

Two reasons here. First, Class A was professional racing, so it was supposed to be wide open. Second, although singles were used mostly in the short tracks, and some of the pro hillclimbers used the inline fours from Indian, Ace or Excelsior, the big fours were, well, big. They might have worked on the hills but they had no chance on the half miles or TTs and so on.

As a side influence, the side-valve Indian racers, two carbs and all, were usually as fast as the ohv Harleys. Practice hadn't caught up with theory yet.

But. During this period, say the late teens through the twenties and almost into the thirties, motorcycle

Peak of an era, the eight-valve Harley-Davidson. There was minimum front suspension and solid rear frame, with no brakes. Oil and gas were carried in the saddle tanks, and the pump on the left side delivered oil directly to the crankcase. It had one speed and a gearing so high the engines had to be towed to start. Harley-Davidson

racing was really neat. It was fast and thrilling and technology grew almost before your eyes. There were several strong companies—Harley and Indian and Excelsior to name three—and they fielded strong teams and put on good shows.

Then came the lean years. Being fair here, the decline wasn't due only to economics. Racing became more and more dangerous and speeds exceeded the limits of the tracks, especially the banked board tracks.

The racing teams were as exclusive as they were strong, and it came to be that if you weren't on one of the teams, you had virtually no shot at winning a professional race. You wouldn't have the money to build a competitive bike and the factory wouldn't sell you one.

Plus, the racing machines of that day were ohv singles (known as Pea Shooters because that's sort of what they sounded like), or ohv twins, ioe or multiple valve. They were special made and not sold in stores, as they say. The motorcycle owner or rider then was likely to go to the races on his side-valve 45 or 74 cu. in. twin or, if he had the money, an inline four. What he rode wasn't much like what he saw being raced.

Thus, the Depression magnified the powers working against motorcycle racing. There were fewer people with the money to buy a ticket to the races, even if they'd been interested. Motorcycle sales declined, so the factories went out of business or cut back on expenses. And the racing team was a popular place to start cutting, because nobody's ever been able to prove that racing sells machines. Professional racing declined to a handful of national events, with only a few riders likely to win, against weak opposition.

Something had to be done.

And it was.

But first, a few comments. Those early racing motorcycles were fabulous and amazing, and obviously they set the ground rules for what was to come. They are worth studying. Stephen Wright's *American Racer* and Jerry Hatfield's *American Racing Motorcycles* are both excellent accounts of those days.

Obvious similarities aside, in general, racing motorcycles before 1930 didn't have much to do with racing motorcycles since that time, nor with the Harley-Davidsons built and sold during that time. However, the racing Harleys of that period do have quite a bit to do with the production Harleys from back then and until now.

So, we begin our study with one of the most basic and all-encompassing changes ever made in racing, a total reversal of the rules, and the adoption of what we'd now call Showroom Stock.

The road racing version of the 21 cu. in. single. In this form, with rear brake, fenders and hand shift, it was for sports or sporting events. The same engine was modified into the oval track Pea Shooter used for Class A racing in the twenties and thirties. Harley-Davidson

When imports revived short track, or speedway, Harley made this CAC—a 500 cc 30.50 cu. in. ohv single. The weight distribution, steering geometry and gearing make this 1934 model almost identical to the bikes used in speedway 50 years later. Harley-Davidson

Chapter 1

The birth of showroom stock (1934-51)

E. C. Smith was the executive secretary of the AMA. He was also nobody's fool. By 1933, the 200 American motorcycle makers had dwindled to two, Harley and Indian. The flock of motorcycle magazines had been reduced to one, *The Motorcyclist*, and that was still in print because the AMA had bailed it out, just as the factories had saved the AMA.

Smith thus had a pulpit, so to speak, and the power to influence, even impose, racing rules. When he spoke, he was a man to be listened to.

In 1933, in the pages of *The Motorcyclist*, Smith called for change. The rules then in effect allowed full racing engines, 21 cu. in. singles for circuits and 750 cc machines for hillclimbs in the professional (Class A) events, and 45 (750 cc) or 80 cu. in. ioe engines for the semi-pros (Class B). The machines were too expensive and they stifled development, Smith said; something should be done.

Sure enough, something was. The nomenclature was awkward and foolish, but the reasoning was sound and the new rules were both fair and effective.

Name first. Remember, the racers were classified as A, full pro and the top guys, and B, not quite into the first cabin yet. They rode Class A machines, limited only in displacement.

A new rule designation was created: Class C (as in A,B,C) and the C riders were supposed to be there for sport. They would be riding Class C machines (we'd call them Showroom Stock).

Class C motorcycles had to be produced and cataloged by a factory and offered for sale to the general public for use by the public on public roads— no competition machines allowed. Production was defined as at least twenty-five examples of a machine having been produced before that model could be considered eligible.

The manufacturer was required to file a detailed list of specifications, such as bore and stroke, wheelbase and type of various materials, right down to the last nut and bolt.

Next, the rider of the machine was required to own that machine and have the papers to prove it. When the rules were first announced, they specified that the rider had to ride the bike to the event, where he could remove the road equipment not allowed on the track, then race, then put all the lights and such back on and ride the thing home. Some guys did it but not many and the rule didn't last long. But you can see from this the intention of the idea.

This was clearly an Olympian concept. It was a way to get the factories out of racing, a move they didn't mind being forced to make. Unfortunately, this worked out like all Olympian systems do—not quite the way the creators had in mind. We had the heavy-handed suggestion, then the framework of the rules.

And we had a name, in my opinion, a poor name. Because the professional racers and purebred machines were Class A, and the semi-pro guys were Class B, the new class, of stock machines and sporting racers, was called Class C. It was logical to those in the know.

The fan, though, somebody new to the sport, only knew that A was for honor students, B meant Dad didn't take back the car keys and C was, well, average. Third class. And I believe the dull name has hurt professional racing for fifty years.

Be that as it may, the AMA guys were smart enough not to impose the new rules all at once. At the time Class C was announced, there were Class A races already scheduled and they continued to be held until much later.

Nomenclature
At the time our story officially begins (the early 1930s), Harley-Davidson offered for public sale and use on public roads the Model D, a 45 cu. in. V-twin;

the DL, the same machine but with higher compression ratio; and the DLD, with larger carburetor and compression ratio, the sports solo.

This was the basic Harley, and the outline will be with us for some time, so I'll spell out some details. Harley-Davidson has a standard code: The single letter designates the basic model, in the DLD's case, D; L stands for higher (why L isn't for lower has never been explained); and the third letter repeats the first or is sometimes an H (again, no reason given).

The D, DL and DLD shared an included angle of 45 deg. between the cylinders, as Harleys almost always do. Bore and stroke were 2.75x3.8125 in., for a displacement of 45.32 cu. in. There were four cams, four shafts with one lobe each, housed on the right side of the engines. The valves were in the cylinders, next to the bore, or side valves.

Harley-Davidsons don't have crankshafts. According to the parts books there are sets of flywheels, one left and one right. These two flywheels are pressed together on a pin, the crankpin, and on the pin are two connecting rods. One rod, the front one in most engines, has a narrow lower end and the other rod has a bifurcated lower end, and the narrow end fits between the two halves. They used to be distinguished as male and female, but now it's probably safer to call them fork and blade.

This assembly rode inside a set of engine cases. (Plural here, although one would think the left side and right side would combine into *a* case, like the crankcase of a car engine. But they don't. The assembly is always referred to as the cases, awkward as it sounds.) The cases were aluminum, the cylinder barrels and heads were iron, and the flywheels were iron in the D and DL and steel in the DLD. Pistons were called Dow metal, an early form of magnesium.

Next, we need to coin a phrase and say the D was a *modular engine*. Back then, motorcycle engines (as well as car engines) were made as separate assemblies; that is, the engine bolted to a flywheel housing or bell housing and the gearbox bolted to the other end. So there was an engine and a transmission, and they came apart. Nobody considered this odd or backward and cars came this way for another forty years or more. But when the motorcycles began to have the engine, primary drive and gearbox all in the same package, the same box, it was called *unit construction*. We don't have the word un-unit, so I'll use *modular*.

Types of racing

In the early thirties, professional racing had narrowed to three venues. First and most traditional was the flat track format; an oval with prepared dirt surface, generally at a fairgrounds. The pros by this time were limited to 21 cu. in. singles and were going faster with them than had the early 61s.

Then there were the professional hillclimbs. These were easy to put on, as all you needed was a hill and permission from the farmer who owned it. It made for a great show, and because speeds were slow it was safer than it looked. Then, as now, some people used the largest engines they could find, but the professionals were limited to 45 cu. in., 750 cc, with no other limits.

A European type of racing known as speedway was popular in some places, having been imported from Europe and ably promoted. The tracks were usually even tighter than the fairgrounds ovals, and the bikes, also imported in most cases, were 500 cc 30.50 cu. in. singles.

There were other less formal racing venues. A few board tracks had survived although most of them were converted into a sort of show, not far removed from the Wall of Death, in which the riders and promoter knew well in advance who was going to win each night's race.

There were only a few road races and those were generally held on borrowed public roads, paved or dirt. There were a few TT courses, with grandstands

The ideal personified: This is Griff Kathcart, on a WLDR at the Wisconsin State Fair in 1937. Notice the license plate mounted sideways on the rear fender, and the hollow shell of what used to be the front brake. Kathcart has done exactly what the AMA hoped he'd do: He has stripped a road-licensed production bike and taken it to the races. Harley-Davidson used to send a photographer to the races every year to record the Harley riders, sort of like school pictures. So we get a good idea of how typical racers looked, although the records don't show how Kathcart did in the race that followed the photo session. *Harley-Davidson*

as well as jumps, but as many TT races were held on borrowed land marked for the occasion. Odd though this sounds, with motocross being the closest relative to TT now, back then the bikes were big and the courses rough, so rough that speeds were low and engine power wasn't important. Because of that, and because a TT was easy to stage and amateurs could do it, the TT classes were 45 cu. in. or open, mostly 80 cu. in. for later 61s. The big ohv twins and the bigger flatheads from Harley and Indian competed over the jumps and did better than a modern fan would credit them.

Bad times or not, Harley-Davidson was still in the racing business in the early thirties and made machines for all the above types of racing. The best known was something of a forerunner to the Showroom Stock Class.

Although the American buyer much preferred the largest V-twin affordable, Harley has periodically offered a single-cylinder motorcycle. During the period from, say, 1926 through 1934, the single was one basic set of engine cases with a choice of attachments and tops: The 21 cu. in. engine came with side valves or overhead valves, magneto ignition or a generator and battery-powered points and coil ignition. The road model had leading-link front suspension, three-speed, hand-shift gearbox, rear brake, fenders, sprung seat and more, while the racing Pea Shooter had no brakes, no fender—except what the rider sat on—and rigid suspension.

The speedway model was the CAC, also rigid and of course with magneto and one speed. It was loosely based on a road machine, the side-valve 30.50 single.

The hillclimb Harley was an oddity that can't be explained by anything the factory has done since.

In 1928 or 1929 Harley released a small fleet of hillclimbers, loaned to a select few Class A riders. The machine had a long wheelbase and non-Harley forks, and the engine was a 45, the limit for professional hillclimbs, but had overhead valves, kind of like the Pea Shooter in design. There were two valves per cylinder but the exhaust valve (singular) had two ports, one on each side. One might suspect this was done so people would think the engine was a four-valve but research showed that if you didn't know how to shape the port correctly, two poor ports were better than one. A few years later the factory re-called the fleet to fit the engines into new and different frames, with front suspension like that on the road models, and then loaned the bikes out again. They remained in professional competition, some of them anyway, for another twenty years.

The official name for the model, or the engine, was DAH. That should mean this was a version of the D, but it wasn't. Same general idea, same displacement, but unlike the road and racing 21s, the D and DAH shared no parts whatsoever.

Furthermore, the factory's photo collection shows a different version of the DAH, one fitted with fenders and with a frame like that of the D, and with four full-length exhaust pipes, the left pair high and the right pair low to clear the kick start. This version has brakes front and rear. That photo was taken outside the plant in Milwaukee, yet the historical record says this was a motorcycle made and sold for road racing in Europe. Harleys did well in Europe, England mostly, in the teens and twenties but they don't show up at all in the thirties.

Interesting, and a bit puzzling. If nothing else, it illustrates first of all, the way the factory built what was needed and, second, that the machines came into and out of existence and were loaned out and retrieved, or not.

Class C

Class C racing was built, carefully and slowly, atop the structure then in place. Class C events were organized for hillclimbs, flat tracks, TT and the occasional road race. And the races were run parallel to the Class A championships.

Here's the important part, one that will stay with us for another thirty-five years: Class C events were to be held for, and contested by, stock motorcycles (production motorcycles) that displaced 45 cu. in. if they were side valve, or 30.50 cu. in. if they were overhead valve.

The criticism of the rules wasn't fair, or at least was overdone. Even if one assumes that Harley and Indian controlled the AMA at the time, and that they weren't going to write rules that would hurt their own interests (which they weren't), there were still reasons to keep the original limits.

Mostly, the rules called for 45s because Harley and Indian both were making 45s, side-valve V-twins of the same general design. Both had rigid rear frames and front suspension, the engines were close in output and the machines were close in weight. The Harley shifted on the left and had the throttle on the right. The Indian was the other way around, and partisans of each would spend days heaping abuse on the other, despite—or because of—their parity.

At the time, the Japanese motorcycle industry consisted of Harley-Davidsons made there under license. Later, the Japanese military establishment ran the foreigners out and gave the license to a firm that made Harley copies, but that's another story. The British, Germans, Italians and others had thriving (well, healthy anyway, considering the worldwide Depression) motorcycle industries, but they weren't selling in the United States. There were only one or two importers of foreign motorcycles in the country, and you had to hunt them down and persuade them if you wanted a foreign bike. Even though there was only a handful, sometimes they were raced, so the AMA decided it would only be fair to let them run. The displacement limit seemed fair at the time, so that's how the framework was laid for Class C racing.

As a backward defense of this issue, one can admit that there was in fact some injustice done. Because TT was considered a venue where power didn't rule, and because TT racing was popular with the sporting crowd, the rules allowed larger engines for such events. Class C had the same allowance, with an Open class. Open here meant 80 cu. in. for side valves and 61 cu. in. for ohv engines, and that in turn was done because Harley was making 74 and later 80 cu. in. side-valve twins and Indian had an 80. Both were threatened by the earlier Harley JD, an ioe engine. That was made as a 61 and a 74, and private owners plagued the factory after the J engine was dropped in favor of the cheaper VL. The Js used to beat the flatheads in races, so the rules put the older engines in their place. (Just thought you'd like to know that. Keep it in mind for when there *is* some injustice, which there will be.)

At any rate, while there were no national championship races for Class C yet, there were events for the class and private owners did enter and enjoy themselves and draw crowds. So the Harley factory phased out the hillclimbers and the singles, both 21 and 30.50 versions.

W series

Meanwhile, the road-going Harleys evolved, too. The D series became the R series, with the generator laid down in front of the front cylinder where it used to stand up, giving people the impression the engine was a triple. In 1937 the R was replaced by the W series.

Replaced is the word here because while the engine had the same bore and stroke and general configuration, it used a new oiling system, one first seen on a Harley with the ohv E series introduced in 1936.

Previously, Harley and most motorcycle engines used lubrication provided under the name total loss; oil was pumped by the rider from a tank into the crankcase. When that oil had been used (either burned or leaked away), a stroke or two of the pump replenished the supply. This worked just fine when engines were lightly stressed and weren't revved

TT/Steeplechase rules allowed larger engines in Class C. This is Hollister, California, 1937, and Len Andres, left, is collecting his trophy for the 750 class won the previous day on a stripped WL—not the equipped one he's on here. Presenting the trophies to winner Jack Cotrell (on a Harley EL Knucklehead) are Captain Ben Torres and Pat Spear. (In those days, the Highway Patrol promoted races.) On No. 7, another EL, is Jerry McKay, and standing next to him is Dud Perkins, a pioneer Harley dealer and racer and, later, sponsor. Len Andres

beyond a few thousand rpm. Power equals heat, so when there wasn't much power there wasn't excess heat for the oil to carry away from the metal bits.

But as power and speed increased, the factories switched to what's now called a dry sump, which provided for oil in the crankcase and much more oil in a separate tank. A double pump or sometimes dual pumps pushed the oil through the engine and back to the tank, and pulled oil from the tank into the crankcase.

The W series of Harley 45s had this separate tank and double system, pressure and scavenge. Otherwise it was a normal sort of Harley, with four cams on the side, three-speed transmission with hand shift and foot clutch, drum brakes, lights powered by battery and 6-volt generator. The new models followed the factory coding system. There was the plain ol' W, the WL with higher compression ratio and the WLD, the sports solo.

There was some confusion, or there will be, because in 1937 there was a model called the WLDR, with special aluminum heads. Then, in 1939, the same model was renamed the WLDD. In 1940 the letters were changed again, to WLD. This was because in that year the factory broke part of the ice with a semi-

Sam Arena, probably the best of the Harley Class C racers in the rule's early days. This was taken after Arena set a new record for the 200 mile national at the Oakland (California) Speedway. Riding this WLDR tuned by Tom Sifton, Arena took 19 min. 20.6 sec. off the old time for the race. Arena also won in Class A bikes and in hillclimbs. Harley-Davidson

The Velodrome, San Jose, California, 1937. Races on bicycle tracks weren't far removed from the thrill show Wall of Death, so riders and promoters usually arranged the winner beforehand. Winning here is Len Andres on a Harley RLD. Len Andres

racer, the WLDR tag now tied to a model that had road gear, such as lights and brakes, but which came from the factory with the engine modified, with hot cams and the like for speed.

Nor was it a season too soon. The last big-engine Class A championship race was held at the mile track in Syracuse, New York, in 1938. The first national championships for Class C were held on dirt tracks in 1939. The class quickly proved to be popular and Harley and Indian management presumably wondered if it wasn't time to get back into racing.

WR

Thus, for the 1941 model year, Harley brought out the WR. It was a pure racer, nothing to be removed. And although the W in the name meant it shared parts with the road machines, it was a bunch different, too. Harley-Davidson went a long way with the WR.

Why? In large measure because Indian's Sport Scout was lighter than the WLD, and was faster. Indian's Ed Kretz was doing in the 45 class what Harley's

Joe Petrali was doing in the 21 class, and H-D management wanted to let the customers begin with more of a racer than they could build from the WLD.

Still another factor here is that the WR was less like a WL or WLD than the factory or the press ever

The WR, here a 1941 model with some later bits, evolved out of the WLDR and came to have a lighter frame, spool hubs, no brakes and foot pegs instead of floor boards. Bill Hoecker

talked about. This is a natural consequence of a production-based formula: If you base the class' appeal on the fans' belief that they're watching what they own, you may take away from that appeal if the fans know it's not quite so. Or so the establishment seems to have believed.

A third reason for developing the WR was that by this time, Harley-Davidson had a racing department, under the direction of Hank Syvertson. There were no factory-paid (salaried) riders. Instead, the racing shop built bikes for customers, gave help to those whom they thought deserved it and loaned out machines for special occasions.

Because the WR series began as the production WL, the engine was of the same basic design, same bore and stroke, and the assembly was the modular type; the crankcases were attached to the primary case and the gearbox was attached to the other end of the primary case, and both bolted to a frame.

Here the intricacies begin. Class C, and thus the WR series, was organized around two different forms of racing. There was flat track on the fairgrounds ovals, and there was TT and road racing. Flat track came out of the board and banked oval era when the rules prohibited brakes. Yes, that sounds reckless and

The WR first appeared in 1941. This is in TT form, with the small fuel tank and oil tank side by side on the frame backbone. It's a 1950 version, with the magneto atop the timing case. Notice the fenders and brakes, and the seat post sliding in the rear frame upright. Harley-Davidson

risky but it wasn't, not nearly as much as it sounds. But while in the pioneer days of Class C, riders were expected to ride to the track, remove their brakes as well as the lights, then race, then bolt the stuff back on, in truth nobody did it, at least not for long. So there was no point in selling the racers brakes, so the WR didn't have any.

Flat track was raced on smooth surfaces, on treated dirt. The machines didn't have much power and they couldn't chop the surface, so the tracks were smooth and thus there wasn't much stress; so again, they could be lighter than road machines.

WRTT

What happened very quickly was two—yes, two—versions of the production racer. There was the WR for flat track, and the WRTT (remember, TT was more like road racing with a jump) for TT and road racing.

The WL used a single-loop frame, full cradle. Frame tubes encompassed the engine, which mounted within the triangle formed by the frame. The WL frame had a steering head of cast iron and a rear engine mount of cast iron, the two joined by steel tubing.

The WRTT had the WL frame with various lugs removed, while the favored team guys got special chrome-moly frames.

The WR used a different frame, same cast steering head and gearbox mount but with the steel tubes

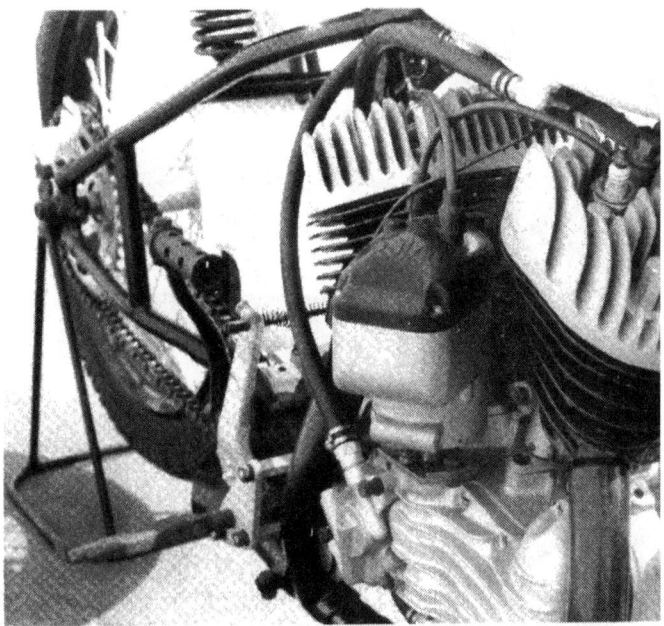

The WR had aluminum heads and cases, and iron barrels. The Wico magneto was originally made for a four-cylinder tractor (notice the alternate blanks in the magneto's cap), and was adapted because it was cheaper than having a magneto designed and built for this limited application. Behind the mag was the line from tank to oil pump, then the clutch lever, then the drilled kick-start lever.

The WR's hand shift came straight from the W series of road bikes, while the clutch pedal was from the three-wheeled Servi-car. The front and rear suspension was like those of the road machines but this machine has the later light frame, introduced in 1948. The WR was 100 lb. lighter than the WLD.

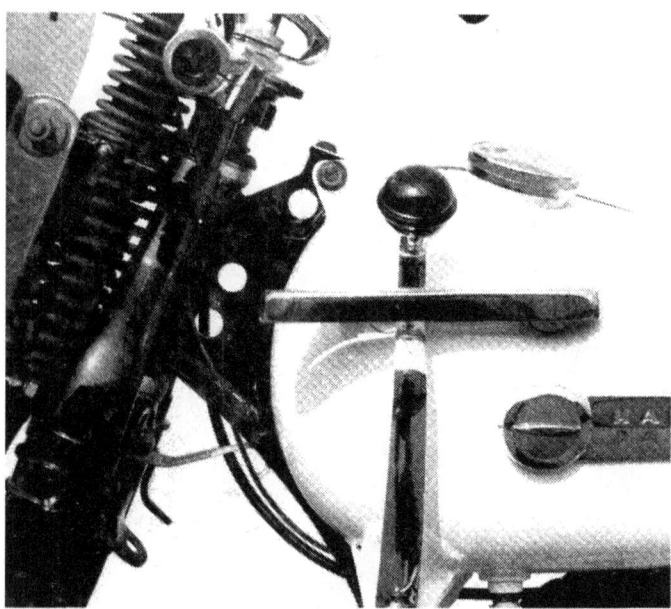

Hand shift used detents to position the lever. The cast steering head was drilled at the factory. The sliders with the knob on the left were a friction damper, sort of a proto-shock absorber, although, in normal Harley fashion, the factory had its own name, Ride Control, for the device.

Crankcases were exactly that—two halves of a housing for the flywheels. In the center of the cases was the left drive shaft, or main shaft, over which went the engine sprocket. The cases attached to the frame and to the primary case. The drilled rods were standard. These are new cases, found on a dealer's shelf by a lucky collector.

made of chromoly tubing, stronger and lighter than the mild steel of the WL frame.

The WRTT had a solid seat mounted on a sprung seat post; the WR had a seat with springs, solidly mounted to the frame. The TT had floorboards and WL pedals for brake and clutch; the WR had pegs, no foot brake, and the clutch pedal from the three-wheeled Servi-car.

Both racing machines had the springer forks, the leading-link front end used by the WL. This was a good system; its two inches of wheel travel saved the rider from the worst of the road.

The WR had vertical braces welded into the rear frame extensions just ahead of the rear hub, but that was the only place the WR had more of anything than the WRTT. With what the parts book and the racers called the "light" frame, the WR weighed 12 lb. less than the WRTT at the factory's loading dock.

Limitations

WR and WRTT had the same engine, very different from the plain W version, albeit most of the differences were on the inside.

Side-valve engines were the norm in 1940; the Ford V-8 made sure of that. Like the legendary Ford, the Harley WR became a great racing engine only because clever and determined engineers did wonderful work on what was essentially a poor beginning.

The valves were next to the cylinders, with the ports and passages and valve seats part of the cylinder casting. The Harley engine had the two crankcase halves bolted together, with the flywheels inside. On the rider's right, inside the cavity cast into the right crankcase half, were the four cams, set fore and aft and driven by gears in turn driven by the right side shaft, known as the gearshaft, that ran from the

WR barrel (left) had larger valves set closer to the cylinder bore and tipped slightly toward the bore. The area around the valves and between them and the bore was shaped, ground away and contoured, relieved in tuners' language, while the WL barrel (right) had no such shape. WR used a tacky coat of aluminum paint in place of a head gasket.

right flywheel through the main bearing and into the timing case.

From the front the cams were exhaust, intake, intake, exhaust, with the exhaust ports at the right outer edge of the vee and the intakes in the center of the vee. Intake pipes (a manifold of sorts) went to a single carburetor mounted on the left side of the vee. The valve stems ran through collapsible covers and met valve followers riding in the case half.

The limitations inherent in this valve configuration are obvious. The best thing you can say about the routing of the intake and exhaust gases is—tortuous. Because the charges must go above the valves and across the span from valve seat to cylinder wall, compression ratio is limited; squeeze the charge and you restrict the flow. And the combustion chamber is bound to be in the wrong place, above the valves when it should be above the piston.

In sum, the only thing working in favor of Harley's side-valve engine was that the chaps at Indian, at least as smart and energetic, had the same basic flaws to deal with.

Innovations

Harley began building the racing engine with the flywheels, steel instead of cast iron. The crankpin—the part holding the flywheels together and with the mated connecting rods attached—was 1.25 in. in diameter, while the W crankpin was one inch. WR main bearings were ball rather than roller, and the camshafts rode in ball bearings rather than bushings.

The W engine had its generator in front of the front cylinder, mounted to the timing case and parallel to the ground, horizontal. The W ignition was what Harley called a *timer*, a set of points in a movable

The WR's valve lifters, or followers, were also called shoes, presumably because of its shape at the bottom. The shoe-shaped surface, at the right, rode on the cam lobe. At the top left was a threaded shaft with a cup that snugged up to the valve stem. The shaft had flats, and there was a lock nut between the housing and cup; that's how valve clearance is set. The follower ran in the guide, at top, and the guide slid into the case and bolted into place. All the moving parts were renewable.

On the engine's right side, the timing side, the main shaft came through the right-side main bearing, a ball bearing in the WR's case. Gears on the shaft drove the four cams, the sequenced front exhaust, front intake, rear intake and rear exhaust, and also drove the oil pump, at the right rear of the cases, and the magneto, either atop the timing case or on the front of the case. In this photo, the rear intake cam and shaft are not quite in place. The shafts turned in ball bearings, set in the case inboard and in the timing cover outboard.

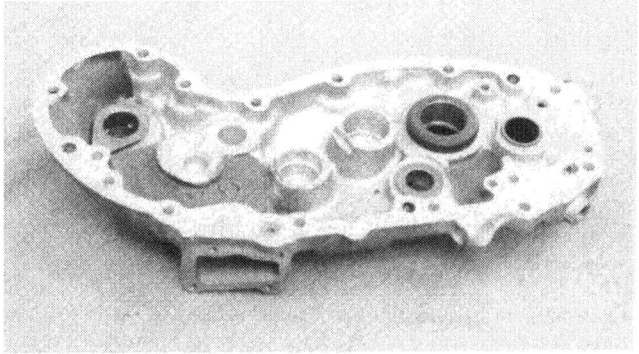

Inside the timing cover, this early example had cast housings for the cam and mag drive bearings, and on the bottom, mounting for a breather, to collect oil draining from the timing case and crankcase.

casing, atop the timing case and driven by the same train of gears that drove the cams, generator and oil pump, which bolted to the bottom of the timing case.

The WR engine had no generator, so it first came with magneto ignition, an Edison-Splitdorf, fitted horizontally where the generator was located on the W engine. In 1948 the timing case cover was changed and the magneto, a Wico, went to where the timer used to be. From 1950 on, the buyer could have the magneto in front or on top.

Several points here. First, the magneto gave a stronger spark and had fewer moving parts than the timer with coil. Next, there was no official word, but apparently the mag was moved from in front of to on top of the timing case so it could rotate back and forth and set the timing. The original mounting was fixed. You engaged the gears and slipped the magneto into its mounts and that's where it stayed. When it went atop the timing case, it could easily be timed. And when it went back onto the front of the case, there was an access plate on the case cover so the timing could be changed or adjusted from there, with the magneto in place.

This won't be the last time magnetos are mentioned. The units fitted in those early days were obtained from a variety of sources. They often were adapted from other applications, tractors for instance, and were never meant to be raced. So the factory used Edison models, mounted vertically, in 1941, '46 and '47. In 1948 and '49 they had Wico magnetos vertically sited or Edisons horizontal. And in 1950 and 1951 you could get a Wico in either place, with vernier adjustment for horizontal only.

Adding to that, the Harley-Davidson and any such V-twin has an irregular or staggered firing order. That's what gives it the distinct exhaust beat and (some say; none can prove it) why the V-twin works so well in flat-track racing. Each cylinder fires every two revolutions. They alternate, with a power pulse every 360 degrees plus and then minus the angle of the vee, in this case forty-five degrees. So the camshaft inside the magneto has two lobes, shaped irregularly to provide this staggered timing. The coils, also part of the

Early timing case cover had horizontal mag drive in front, right here, and oil pump drive at rear.

Later WR timing case covers were shaped closer to the cases and provided for mag drive at the top of the cover. There was an access plate, at the right here, behind the exhaust pipe. With the plate off, the mag drive could be disconnected and timing adjusted with the magneto in place.

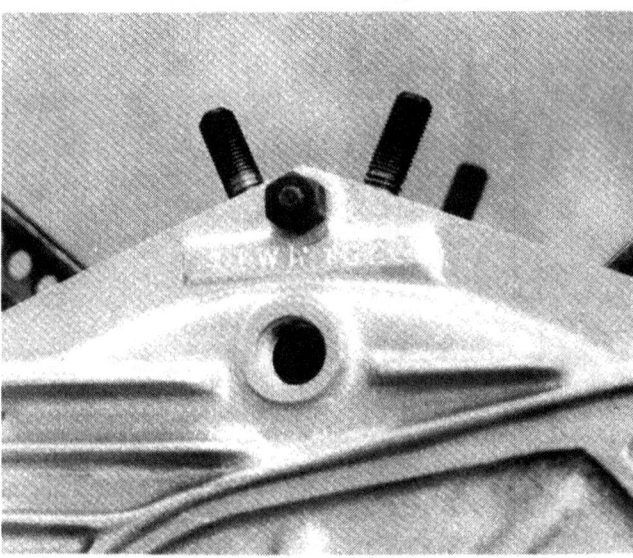

Nice, neat numbers on this crankcase were possible because the cases weren't numbered when they were found in the back of a dealership.

magneto, are called upon to deliver their sparks in the same staggered sequence. This all adds up to an ignition system that worked fine, well, as well as the alternatives of the day, but which will present serious problems.

The WR engine differed most from the W or WL where it showed the least. The WR cylinder barrel castings were different from, and had ports larger than, the W barrels. The WR used different cams, as in lift and timing. The cams were also different in that they were ground for either flat followers or roller followers. The rollers were supposed to follow the contours of the cams more accurately, while the plain followers were lighter and were supposed to allow the engine to rev faster before the valves floated; that is, they no longer went up and down in time with the cams.

The W engine, along with most side-valve engines, had the valve stems parallel to the bores. The cams and gears rotated in the same plane as the flywheels, engine and clutch sprockets, with everything at right angles.

The WR engine didn't work like that. Instead, the valves were convergent to the bores, tipped toward the inside. The valve seats were tipped this few degrees, and the bores for the cam followers in the right case half were angled as well.

Why? Remember the twisted path the intake and exhaust ports trace. The gases must make a full 180 deg. turn in the W, but with the angled valves and

Standard engine sprocket for the WR was at left, with its finer and smaller teeth. The sprocket, at right, was for the WRTT in long races, where the smaller, lighter chain could stretch beyond endurance. The optional sprocket used the same primary chain the 61 and 74 big twins used.

seats, the path is somewhat straighter. Enough so the engineers must have had to come up with proof of the benefits or they wouldn't have been allowed to impose the difficulties this angling entails.

Those who have followed this technical complexity should say at this point, Hey! If the valves are angled toward the bore, then the valve stems and/or

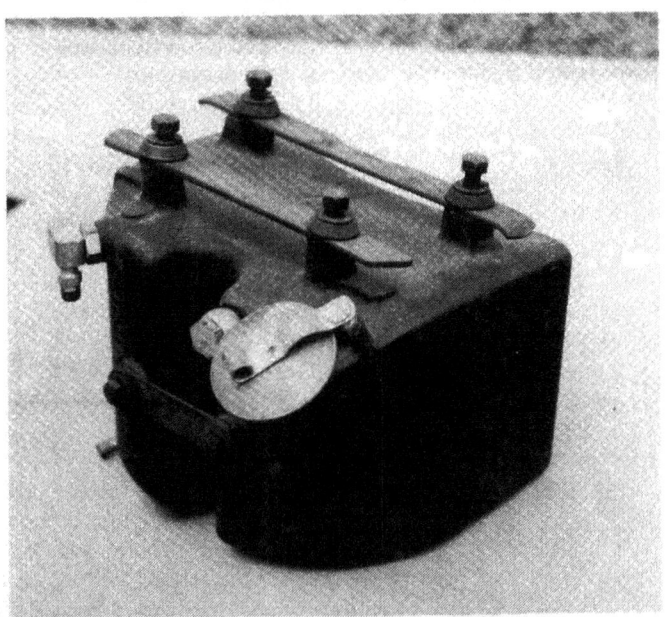

This oil tank was for the GT version of the WRTT: For long races, like Daytona or Langhorne, the WRTT had double fuel tanks and carried five gallons of gas. This tank attached to the frame, behind and below the seat. The bars at the top straddled the frame tubes. The filler cap, front and center here, was quick-release.

Rear hub carrier for the WR frame was a casting, welded to chromoly steel frame tubes. Because there was no rear brake, there was room for a different ratio sprocket on the other side. This was later outlawed.

the followers aren't at right angles to the cams, right? So, how can the cams work at an angle?

They don't. The followers, either roller or flat tappet, have their lower surfaces ground at the angle by which they're tipped, squaring things up. The followers, known as *shoes* by the builders of these engines, must be carefully checked so they're installed right. The tipped valves aren't common knowledge, and there have been mechanics who put the followers in backward, resulting in damage to the cams and followers.

The engine speed parts weren't all that different, in fact the factory offered a tuning kit for road machines. For $109 the owner got cylinders, heads, cams, tappets and an intake pipe. Buy or supply the carb and exhaust pipes and the home builder had a fair approximation of a WR engine. Or it could be built by the dealer as a hot road bike version of the racer. Harley kept this practice with the later KK and KHK versions of the side-valve K and KH road bikes and with the original hot Sportster, the XLC.

AMA influence

And then came all the other parts; a direct result of the rules. The intent of Class C was pure stock, showroom stock, so while the owner was allowed to change the parts he got, as in milling the heads or drilling lightening holes in the pedals, he wasn't allowed to swap parts. As always, the racers objected and in truth they did need some changes for some races.

So the AMA allowed swaps, but the parts exchanged had to be recognized, official production factory parts. There was a choice of 18 or 19 in. wheels, with 3.25 or 4.00 in. tires. There were solid or rubber mounts for the handlebars. The WR had a dual tank, with gas on the left side and oil on the right. The WR tank held 1.74 gal. of gas, plenty for most dirt-track events. But the WRTT had an option: the wide

---

Joe Petrali: The unequaled record

Because every season is a whole new ballgame, so to speak, with different rivals and often different rules, it's never been quite fair to compare racers from different eras.

Even so, while it might not be fair, it's safe to say Joe Petrali's record will never be equaled, much less beaten. In 1935 he won every national championship race, all thirteen of them.

Petrali was the consummate racer of his day. Born in San Francisco in 1904, he rode his first motorcycle at age twelve, went to work at the local dealership at thirteen, won his first event, an economy run, at fourteen and had a borrowed Indian at sixteen, when he went up against the all-conquering Harley team on the Fresno board track in 1920. They boxed him in, but Petrali got loose and finished second. For his next big race he had a team Harley, borrowed from Ralph Hepburn after Hepburn broke his hand. Petrali won, doing the 100 miles in twelve minutes and some seconds less than an hour. After that, Joe was on the team.

He was a team player. When Harley dropped out of racing, Petrali rode for Excelsior. Then when that company left the motorcycle business, he switched back to Harley just in time for the Depression, and in time to become, in effect, *the* Harley racing team when it was the only factory team. There was no official points championship then, but Petrali was high-point man in 1931, 1932, 1933, 1935 and 1936, as well as national hillclimb champion for eight years in a row, 1929 through 1937.

Obviously a talented man, Petrali is best known for his 1936 ride on the streamlined Model E, the just introduced Harley big twin of 1937. That machine was clocked at 136.183 mph at Daytona Beach, an American speed record and a world record for unsupercharged motorcycles. The machine, at this writing, is on display at the Smithsonian Institution. Thing is, the sleek shell in which the machine is wrapped, for display and for photos of the record attempt, actually contributed to instability at speed, so Petrali's record runs were made with the bike stripped. He's shown here on a Pea Shooter, the 21.35 cu. in. single on which Petrali won all major board-track races in 1925.

Petrali was never at home with the larger, less refined Class C bikes, so he retired from riding in 1938. He went on to work on Indianapolis cars and then for Howard Hughes. Petrali was a flight engineer when the legendary Hughes flew the legendary *Hercules,* also known as the *Spruce Goose,* and after that officiated at Bonneville for United States Auto Club and its predecessors.

tanks known as the Daytona tanks, with 2.5 gal. in each side and a 1 gal. oil tank, an aluminum casting, mounted beneath the seat.

The WR was based on the W series and it was a production racer, so the factory tried hard to keep as many of the parts the same on the WR as on the WL. The parts books even say if the part isn't here, it's because it's a street part and is in the street parts book.

Sometimes they adapted. The early version of the WL used a clutch that had the basket recessed into the hub. This seems to have been a problem for long life, so later models, from 1941 on, had the basket outboard of the sprocket. But racers need to lean over, and the narrow clutch makes a narrower bike, so the WR clutch is the original, narrow design.

Racing may have improved the breed. The first W series frames had a two-bolt mount, fore and aft, while the WR frame came with a gearbox with a three-bolt mount that was used from 1941 on for all W-series frames and gearboxes. (The triangular mount was less willing to rock side to side under torque loads.)

Primary drive for the WR, and offered for the WRTT, was a chain with ⅜ in. pitch, light and as strong as needed. The WRTT had an option for long road races like Daytona: a chain with ½ in. pitch, the primary drive chain used by the 61 and 74 big twin Harleys. There was a wide selection of final-drive ratios (output and rear sprockets), and of internal ratios for the three-speed gearboxes. The several drives for the magneto were listed, as were the flat or roller cam followers. There were alternate camshafts, too, although because they could be machined and reworked, there wasn't as much need for alternates in the book.

Rule bending

This parts supply was, mainly, a good thing, as it let the private racer get the stuff the top guys had. The problem, assuming the rules were followed, was that if listing parts in the book made them legal, then parts that weren't in the book—parts that a factory didn't want or parts for a model that a factory had no interest in—couldn't be used for racing. (This comes later, but the basis for the problem was here, when the rules were devised and first revised.)

I might as well add that these rules were loose enough to allow, perhaps even encourage, creative interpretation and downright cheating.

The racers had plenty of freedom within the framework of the rules. On the left is Ray Andres, with a WR frame and the magneto driven off the front of the timing case. On the right, Paul Albrecht is shown with a WRTT frame and the mag on the top of the case. These shots were taken at the Milwaukee fairgrounds in 1948, by which time license plates and brake housing, with innards removed, were no longer seen. **Harley-Davidson**

For example, lighter chromoly steel was used for the tubes of the WR frame, so there were builders who used the same material to construct lighter and stronger WRTT frames that looked just like stock ones. The steel flywheels from the DLD engine of ten years earlier were stronger and lighter and popped right into the WR cases, so there were racers who used them in what legally were WR engines.

The engines, make that engine cases, got a lot of juggling. First, they were light and not designed originally for racing, so they changed under stress. A good set of cases would cure, that is, flex and then settle into a stable shape. If everything else was equal, a set of cases that set itself permanently in perfect alignment—that is, with the main bearing locations directly opposite each other and perfectly round,

Meanwhile, the boys in the back room were hard at work. Rider is Leo Anthony, and the negatives were found in the basement of Pohlman Studios, the photographers who took all official Harley-Davidson photos for generations. The penciled notes say this bike is a WRL, a designation that appears nowhere else in the Harley archives or records. If the letters followed the code, that would mean a high-compression racing engine, which doesn't make much sense. But, look at the high pipes on the right side, and notice the lack of a gearshift lever and gate on the left side. This is a WR with foot shift! And the frame has two front downtubes and looks to be based on parts from an earlier hillclimber. It appears the engineering department was experimenting here. *Harley-Davidson*

with the two halves of each cylinder mount perfectly true side to side, up and down, fore and aft—would create less friction, have less drag and thus produce more power than cases that didn't line up.

Now. Remember, when the rules came out they required the Class C rider to own the machine. And in those days, frames weren't numbered. Because the licensing requirement (as in, ride it to the races) was soon dropped, proof of ownership became the engine numbers. WR engines were made with no numbers and were stamped later, for convenience. If the engine was going to a dealer or private racer, it would get a normal number. Harley-Davidson used a logical system of model year, model type and sequence number. A customer WR engine, for instance, would read 48WR1234 if it was a WR made in 1948.

The switch here is that favored racers carried with them bills of sale, with an engine number for the machine they owned. The rules didn't allow a rider to swap machines in the event of trouble, so it quickly made sense for the pro with two machines to have both the best bike and the spare fitted with the same engine number.

Thus, the factory didn't always stamp the cases before they were shipped. Years later, lucky restorers have been able to find unstamped, unused cases stored away by racing dealers.

Circumstances thus created some unusual cases, play on words or not.

Teamwork

There was a Harley-Davidson racing department during the WR era, but it was small and mostly con-

cerned with the occasional record run, such as the Petrali streamliner, and with designing and building competition machines for customers, favored or not. The WR and WRTT were in the catalog and for sale to anybody with the money. To illustrate, the 1950 catalog lists standard colors of Brilliant Black, Ruby Red and Riviera Blue, and Metallic Green, Flight Red, Azure Blue or White for a nominal extra charge. In other words, just like the road bikes, or like cars of the day.

This set the stage for what I'll call The Team: a combination of rider and wrench, with sometimes extra help from a sponsor or backer.

News accounts then referred only to the rider, while more recent history credits the wrench—the mechanic usually billed as the tuner—for the results. Nobody ever talks about the folks who made it possible, the people who put up the money.

In truth, it's all vital. The rider and the tuner complement each other and are worthless by themselves. It's not possible to say for sure that Rider A would have won no matter who built his bikes, nor can we know for sure that Tuner A's work would have put anybody into the winner's circle.

It takes both.

The intriguing thing here is how often the two parts of the whole team manage to find each other.

Tom Sifton was a Harley-Davidson dealer in San Jose, California. He was (and still is) a thoughtful, intuitive, methodical man. He began building production racers back in the D-series days, and did well. Like other such men, he enjoyed beating the factory and then rubbing it in, not telling the official racing department how he did it.

Perhaps none of what Sifton did was unique, in that there were other tuners working on Harley engines, and on Indians and on the flathead Ford V-8, all of which benefited from the tricks that made Harleys and Indians fast.

Doesn't look like work, does it? From left, Johnny Butterfield, Billy Huber and Pete Chann, all WR equipped, at the Milwaukee mile, 1947. The rules restricted tire tread to the semi-knobs shown here, while wheel and tires sizes could be anything between 21 and 16 in. The riders here were pros, with factory help but no salaries. Huber is distinguished in the record books: He and Bobby Hill, an Indian rider and two-time national champ, were neck and neck on the Atlanta mile track August 8, 1948, in dust so thick the race had to be stopped. Because visibility was nil the judges couldn't decide who'd been in front when the red flag was waved, so Huber and Hill split the win. It was and still is the only AMA national in which nobody came in second—the man behind Hill and Huber was considered third—and the only dead heat ever recorded. Harley-Davidson

But Sifton thought things out, tried new things and kept careful records. He found that the factory's valve gear wasn't working as intended; the followers weren't following the contours of the lobes and the valves were floating at high speeds. So Sifton improved the springs and contours. He worked out how the cases aged and developed ways to age them and compensate for it: When you compress the barrel, you should torque it as if the head was in place and then align bore it, so you get the true circle you're after.

Breathing isn't just *a* critical factor with a racing side valve, it's the *only* factor. Sifton wasn't the only man to work out that the ports needed to be enlarged and contoured for better flow, or that cutting channels in the cylinder's surface between valve seat and bore was beneficial (the Ford V-8 tuners called this porting and relieving), but he did a top job of using the techniques. Ahead of his time, is probably the term to use here.

Because they used roller and ball bearings, and didn't rev high or have high compression, motorcycle engines from the thirties didn't need fancy oiling systems, witness the retention of total loss oiling.

But the V-twin had some unique characteristics that made oil control and delivery more important than lubrication alone. There were pie-plate flywheels revolving inside the crankcase shaped much like two pie plates bolted together. There was one crankpin with both connecting rods on it and the rods running the pistons up and down, not quite together. There was not a lot of space in the crankcase, while the displacement inside the cylinders below the pistons varied with the displacement above the pistons.

This amiable youth is Joe Weatherly, a good rider who became much better known and a whole lot richer as one of the modern era's pioneer stock car heros. He's shown here at Daytona, which he didn't win, but he did take the road race at Laconia, New Hampshire, and won a half mile in Pennsylvania. His mount here is a WRTT, with the filler cap for the remote oil tank barely visible between the air cleaner and seat. The hand lever on the left grip went to the front brake. (You didn't have to do two things at once using one hand before the hand clutch lever was put on the left.) Daytona International Speedway

This caused compression and expansion inside the engine. If the cases were sealed, they wouldn't stay sealed long under the pressure of two or three atmospheres. So Harleys got timed breathers that opened and closed in sequence with the pistons going up and down. And the engines had scrapers on the flywheels, collecting the oil from the walls of the crankcase and delivering it to the bottom of the cases where it was collected and pumped back to the tank.

The draining and delivery of the oil was a delicate matter and the standard technique in Sifton's day was to oil the cylinders on the turns by shutting off the throttle and letting the vacuum draw oil up past the rings.

The real Sifton story, drawn from Jerry Hatfield's *American Racing Motorcycles*, is that Sifton worked out a better way. He changed the breather timing so the engine got oil on the straights and told his rider not to shut off, to keep the power on in the turns . . .

and went on to break the Class C 200-mile record by nineteen full minutes.

The rider in that epic record-setting race was a man named Sam Arena, the other half of the partnership. And another success story.

Arena began as an apprentice in Sifton's dealership, in the right place at the right time. He raced on his own and then with help. Arena was a winner in Class A and on the ovals, but he easily made the change to the larger C bikes and became the Harley star, in direct competition with Ed Kretz, his counterpart at Indian. When Arena decided to retire from full-time racing, he bought Sifton's dealership and became a hobby racer in hillclimbs, while Sifton went into the cam-grinding business and for fun built engines for the next generation of racers.

Bottom line

Although there were no detailed magazine tests done back then and the factory's official literature is

An odd one. This was taken at Daytona Beach in 1949, and shows a Harley sliding out of the turn with a Norton in hot pursuit. Normal, except the Harley has large telescopic forks, straight from the big Hydra-Glide. Presumably the forks were homologated as an option for the WRTT, although the books don't show it. Daytona International Speedway

deliberatedly vague on details, it's still possible to assemble some specifications for the WR.

Keeping in mind the sources (some of it may be guesswork) it's still safe to say the WR and WRTT were competitive machines in their day. Indian's Sport Scout came close to the WL and WLDR, and so did the occasional 500 cc English bikes that found their way into Class C races. The two rival makes swapped wins all during the thirties and forties, ergo they must have been close in power, power-to-weight and stamina.

Also, the Class C machines got faster as the tuners learned how to make them work and the factories learned what they could do and could get away with.

There was no national championship circuit then, so there was no official points tally. Instead there were several national events for specific distances, tracks and types of machine.

From those results, we can interpret: The 1936 speed record for five miles on a one-mile track for Class A, the 21 cu. in. Pea Shooter singles, was 3:34.6, or close to 82 mph. For Class A 45 cu. in. engines, it was 87 mph, and for Class C 45s, 70 mph. The 21s had slowed down the races on dirt as well as on the boards, and the production-based 45s were slower than the all-out racing 45s, which is just what the rule makers had hoped for. And by 1950, the Class C five-mile record for the mile was 82 mph, headed back toward the pure racing times, as the WR became a pure racer.

Model	WR/WRTT
Year	1941-51
Engine	45° V-twin, iron barrels, side valves
Bore and stroke	2.75x3.1825 in.
Displacement	45 cu. in.
Brake horsepower	38 (est., later units)
Transmission	3 speeds
Wheelbase	57.5 in.
Weight	300-330 lb. (est. dry)
Wheels	19/18 in.
Tires	3.50/4.00
Brakes (factory)	none/drum front and rear

At the beach

This era also saw the beginning of what are now the classic races, such as Daytona Beach. Its sanction began as the road race in Savannah, Georgia, but was moved when that city became too civilized for real road racing, and when Bill France, a mechanic turned racing driver turned promoter, realized that professional racing worked only if it was organized and if it was a good show.

France lived in Daytona Beach, where the beach, the tides, the chamber of commerce and the willingness of northerners to leave home in the winter combined to make that city a center of speed runs, on the beach and usually one vehicle at a time.

That was fine, but France saw more potential for Daytona and so organized car races, on the beach and on the paved road behind the beach proper. To widen the appeal of the place, motorcycles were invited to have their own races, which began there in 1937 and, of course, still continue, albeit in a very different place and form.

The beach races were instantly popular. They were officially road races, even though half the course was sand and there were only two turns, the U-turns at each end. But racers had been coming to the beach since the turn of the century and the promoters were savvy enough to hold the races in the winter, when the rest of the world was snowed in. (Later, when Daytona Beach had become a resort full time, the races were moved to the end of the vacation season, to avoid overcrowding.)

Digressing a bit here, the success of Daytona attracted what few imports there were in the United States. Indian won the first beach race in 1937, then Harley-Davidson with Ben Campagnale in 1938 and 1939 and Babe Tancrede in 1940. Both riders came from Rhode Island, backed by strong clubs and dealers, and with some factory help.

But then, in 1941, a Norton 500 won the beach race and shocked the American club (read here, the two American factories). The 500 singles couldn't be

DAH engine was a 45 deg. V-twin, same bore and stroke as the D or DL production engines, but with overhead valves and no parts in common. Notice that the magneto mounted in front of the front cylinder, a location Harley-Davidson will use for the next 50 years. There's only one exhaust valve for each cylinder but each valve had two ports, thus the four exhaust pipes. This engine was fitted in several road-going frames for research, but never was sold to the public. Bill Hoecker

kept out. But the rules could be modified, and were, with a limit on the compression ratio and a ban (no kidding here) on shifting in dirt track races. Flatheads worked better with low compression, and hand shifts were slower than foot shifts; the American bikes had flathead engines and hand shifts, the English didn't, so...

That's putting things a bit baldly. The Nortons in question were overhead cam models, as raced in Europe and based on the street machines Norton was building at the time.

The problem was, the English were producing racing motorcycles while the Americans were racing production motorcycles. The importers argued that anybody could buy the racing Nortons (which is why Class C began), while the home team argued that the ohc engine wasn't what appeared in the showrooms (also true). At the same time, both sides had their own commercial interests cloaked in this debate over how to interpret the rules, which set the stage for later arguments and revolutions.

Even more of a digression here. Remember that TT bikes were allowed extra displacement but there was a rule in effect keeping the older ohv engines out of contention, in favor of the newer, larger flatheads? Later, when all the flatheads had been retired, the TT rules still allowed larger engines that could run in the flat-track or road events. But the limit became 900 cc, inside which the Harley Sportster fit nicely, and which the Vincent V-twin exceeded. There are those who say this wasn't a coincidence.

Making a living

Other odd stuff from the late thirties: Times were tough, and on occasion promoters saved wear and tear on the equipment and the riders, by working out beforehand who was going to win. This was known as hippodroming, named after the banked board tracks which were in turn named after the Roman stadiums where, history tells us, the fix was perfected, if not invented.

Worse, from the riders' points of view, was that promoters were known to take off with the night's proceeds, leaving the riders unpaid, which left them with no money for food or gas to get to the next race. And there were no factory teams.

The AMA got very tough here. Later, when people had forgotten the abuses, they laughed at the cures still in place—and AMA rules that took away licenses from racers caught taking part in "outlaw," meaning non-AMA, events. Or, they went to court and claimed restraint of trade.

Which it probably was. But, like the unions versus right-to-work laws, the people on one side often hadn't been there when the other side was being formed. For years, promoters were required to post the purse, in case, before a wheel turned on the track. The promoters objected—who wanted to drive around with $20,000 cash in the trunk of his car? But the gruff old codgers who enforced the rule were former hungry young racers who had subsisted on soup made from ketchup, and the rules, By God, were enforced.

In the early days of Class C racing, Harley and Indian generally were equally matched, with Norton the occasional spoiler. Daytona was *the* road race, closely followed (1940) by Laconia, New Hampshire. Springfield, Illinois, was *the* mile. When Class A was officially abandoned, the AMA took a step in the right, long-term direction by creating three rungs in the racing ladder: Novice, Amateur and Expert. The new racer could begin against his equals and work upward, talent and drive willing, by earning points. The system is still in use, albeit the two professional classes are now Junior and Expert.

The top riders could barely earn a living with their winnings, so most of the pros had other jobs or were backed by dealers, or fathers, or sometimes by a man who was both. The factories provided technical help to the top riders and they swapped tips while expecting the same from tuners like Sifton and Len Andres.

Racing skips a beat

World War II (obviously) interferred. Harley-Davidson's 45s were converted again, into the WLA (A for Army), with the strong old flathead getting more beef and less stress and doing exemplary service in all theaters. After the war, there were a number of shops in Europe that made their living converting and maintaining leftover WLAs. Strong, is what they were. And they kept Harley out front in countries where imports were banned, such as England (speaking of old grudges).

When racing resumed road bikes and cars were pretty much like they had been before the war began. The free world's factories had had plenty to do and it took some time to get back into production, much less take advantage of the technical lessons learned from the war.

So the 1947 WR was virtually identical to the 1941 WR. The AMA had changed the hillclimb rules, once more trying to get more people into the sport, and had banned exotic fuels and traction devices. There were WRs rigged for hillclimbs, with two-speed transmissions and chained tires, as well as the flat-track versions and the road-race/TT models. The engines evolved as the tuners kept working on the cams, ports and combustion chambers. Dynamometers of the day weren't always accurate, nor were the stories tuners told each other, but the best WRs, from Sifton or Andres, probably cranked out close to 40 bhp.

Change was on the way. Folklore says that American soldiers in Europe saw the sporting machines and hauled them home. In retrospect, however, it's not likely. There weren't that many Triumph Speed Twins and Norton Manxes and MG-TCs on English roads; they didn't have the gas to burn and the GI

This is a factory photo, taken in 1930, of what has to be a DAH engine but in a road-style frame and with suspension and fuel tanks, even a toolbox on the forks and a rear stand on the rear fender. Except for no lights, this looks like road trim. But the official notes say it's a road racer, built for Europe. Using logic, hindsight and no accountability, it's fair to guess that the engineering department made some road versions of the DAH, in case the market needed an ohv, high-performance model for the 45 class. Times got tough, side valves were cheaper to build and maintain and they worked well enough, so the DAH never got past the hillclimb stage. Or, all this could be the author's overheated imagination.

wasn't all that rich. More likely, as often happens, when the upper crust saw that the middle class was prosperous and buying the best domestic stuff, why, it was time to discover the imports, as in Triumph and MG.

There was also need for transportation, and technical advances had been made. The motorcycle companies knew this, along with knowing that when people have money they're going to spend it on something, so it might as well be motorcycles.

Thus, Harley-Davidson came out with a whole new class of machine, the two-stroke single, called the S-125 and later the Hummer (in 1948). The big twins got telescopic forks in 1949.

Racing was affected only marginally at first. Drag racing and enduros were just coming into vogue, but those events for big or small bikes, were entirely amateur and attracted no factory efforts. Nor were Class C bikes legal there.

Instead, they were good at what they were built for. The WR, Sport Scout and Manx were specialized. The rule books don't reflect this, but later historians have written that both the US factories seem to have tacitly agreed on full race machines and less restricted use of parts. Photos from Daytona Beach in 1950 and 1951 show WRs being raced with Hydra-Glide forks.

As a left-handed reinforcement of this, at the AMA's annual meeting in 1951, a proposal was submitted for the creation of another complete class, Class D. It was for standard road jobs with no parts to be modified or replaced, for novice riders only, and with short-track classes for 125 and 250 cc engines, as well as 750s. And, in 1952 all entries to field meets, field trials and gypsy tours were required to be ridden to the event.

Minutes of the meeting show that there were complaints and concerns. Fewer people were competing in these events and the new and less experienced riders felt they couldn't fairly compete against riders with more backing. Yes, it was the same story all over again. But the ending changed; Class D would never be heard of again. Instead, the rules held still and everything else shifted.

Chapter 2

Reign of the flatheads (1952-69)

When prosperity came home from the war, so to speak, it was hand in hand with a war bride: imports. Bikes, cars, clothing, movies, furniture, just name the field and Americans were having a good time learning how other folks lived.

Harley-Davidson wasn't slow to work this out. Nor did it fail to appreciate that while some of the success of Triumph, BSA, Norton, BMW et al was due to snob appeal, a lot of it was because foreigners made some good machines, with features and advantages the home product lacked.

So Harley got busy. At the top of the line, the big twins got telescopic forks, then foot shift and rear suspension. The top end of the E and F engines got improved porting and oil control and alloy heads. Then the lower end was beefed up to handle the extra power the revised engine could develop. At the less expensive end, the novice end, Harley offered a 125 cc two-stroke, based on a DKW design the Allies had received as war spoils.

In the middle was the Harley-Davidson K, a practical blend of old and new and (in 1952!) the first American import fighter.

The K

Unlike any other Harley in history, the K was almost all new. It had a double-loop engine cradle, where the earlier W series had a single tube around the engine. The K had telescopic forks in front, swing arm with tubular shock absorbers in back (skipping over, thank goodness, experiments with plungers and sliding hubs and other semi-suspensions that failed for other makers). The K had a hand clutch and foot shift, and if anybody ever doubted that the K was supposed to meet and beat the English motorcycles, note that the K shifted on the right. The W, E and F all had hand shift on the left, so the big twins had foot shift on the left, which made sense, while the K had the lever on the right, which made it like the Triumph and BSA.

The K engine was of unit design, before the Brits abandoned crankcases for the primary case and then the gearbox. The K had four cavities in one housing: the cylindrical cavity front and center, for the flywheels held together by the crankpin. There was the cavity on the left center, with engine sprocket, primary chain and clutch. At right center was the timing case, with four one-lobe camshafts, drive for the timer, generator, oil pump and breather. Directly aft of the crankcase was the space for the four-speed gearbox. The cases split vertically, so assembly meant stacking all the inside bits, flywheels and indirect gears inside one half of the case and then popping the other half over the assembly.

The K engine was almost a modern version of the W engine. It had the same bore and stroke, 2.75x3.185 in. It of course was a 45 deg. V-twin, and it had iron barrels on aluminum crankcases, with aluminum heads. And it was a side valve, a flathead.

A few years after the K's introduction, say about the time it was replaced with the ohv Sportster, the flathead engine was generally seen as a mistake. But looking back, for 1952 the side-valve configuration doesn't seem wrong at all.

First, it was a design Harley-Davidson knew how to make and make right. The ohv twins were good engines but they did give more trouble than the flathead 45s and 74s from the same era. As a reinforcement, Indian saw the imports coming and invested all its working capital in a completely new line of bikes, ohv vertical twins like the English made. But the Indians didn't work well, they sold worse, and in 1953, just when the K was going strong, Indian went out of production.

Second, flatheads weren't all that obsolete. Ford, Plymouth, Packard and countless other cars had side-valve engines. They were quiet and easy to service. The big changes, the improvement in combustion chamber design and the increase in compression ratios and thus the real benefit from ohv, were still in the lab or on the dynamometer. The K appeared in 1952, so when the Oldsmobile Rocket V-8—first high-performance production engine from Detroit—hit the roads in 1949, the design for the K probably was complete. Besides, one of several things Harley-Davidson is famous for is not wasting money.

Which brings us to the third point. The 750 cc flathead engine was a match, on the street as well as the track, for the 500 cc ohv engine. (Note: Inch versus centimeter is nothing new but it's still confusing when one leaps back and forth. To keep things clear here, from this point on measurements will be in inch and pound for wheelbase, weight, wheel and tire sizes and so on, but in cubic centimeters, or cc, for displacement.)

Most of the imports came from England and most of them were singles and twins of 250, 350 and 500 cc, so a 750 flathead seemed up to the job. As it may have been except that the English (no fools they) increased their displacements to 650 cc for the sporting twins, so Harley went . . . but that comes later.

The Harley-Davidson K was announced early in 1952. It had a 750 cc engine and was rated at 30 bhp. The K weighed about 400 lb. ready to go, with 4.5 gal. gas tank and 3 qt. oil tank, lights, mufflers, side stand and other equipment.

KR and KRTT

Almost at the same time, came the racing version, the KR, named in keeping with what was by then tradition.

The KR logically was a mix of K and WR, but with some improvements and additions. Most of the latter involved the frame. Both the W and the WR were rigid in back; that is, no suspension except for tires, seat and rider. But the K had rear suspension.

The K frame was different. It had a cast steering head and a cast seat junction (where the single back-

Cream of the crop: Pitched into a classic slide, with rear wheel out and boot serving as an outrigger, is Mert Lawwill, final exponent of the brakeless dirt-racing machine. In hot pursuit are Chuck Palmgren (38), Dick Mann (2) and Cal Rayborn (25). Lawwill's KR used rear suspension and a custom frame. He gives this bike credit for his national championship year, 1969, when the actual design was 18 years old. Harley-Davidson

bone tube met the two rear uprights and where the seat and oil tank attached). At lower rear was the rear engine mount, known in the factory as the lower tomahawk or the twin tomahawk, ostensibly because when the two halves of this assembly are apart, they look like the American Indian's instrument of war.

The twin tomahawk performed several tasks. It held the rear and bottom frame tubes. Near the top of this casting was a horizontal tube through which the swing arm bolt ran, so the casting was the swing arm pivot. There were four passages running fore and aft, carrying bolts for the rear engine mount. And there were lugs at the lower rear.

That's for the K, the road model with rear suspension. Dirt-track machines didn't have rear suspension, so at first, the KR had a frame similar to but lighter and shorter than the K frame. It was made of 0.062 in. wall chromoly steel tubes welded to the cast steering head, seat mount and lower tomahawk. But where the K had shock absorbers and springs running from the seat casting to the middle of the swing arm, the KR had a rear frame, a subframe that bolted to the seat casting and the lower rear casting, and which carried the rear hub solidly in place.

Still using the familiar factory coding system here, there was also the KRTT, announced at the same time, and like the WR and the WRTT, it was the KR for road racing and TT. By this time it was generally accepted that suspension was a good thing for road and TT courses as well as highway riding, so the KRTT had the KR frame except it came with the K's shock absorbers, swing arm, brakes and fenders, but a different seat.

The KR and the KRTT shared an engine and, like the rest of these models, the KR engine was like the K and like the WR, but moreso.

We are deeply into Class C racing here, in spirit and in letter, so the KR engine looked almost exactly like the K engine. Same bore and stroke, same unit construction with the four cavities, alloy cases, iron barrels, alloy heads and so on. The engine mounts were the same, and a KR engine could pop right into a K frame and vice versa. This will be so for the next thirty years, with exceptions to be noted.

The KR was the basic racing model for dirt tracks. The engine lines were new for the K series but their profiles changed little when the K became the X. Important points here include the two-piece frame with single backbone and twin downtubes front and rear, and cast steering head and seat/strut mount for the main frame, with the rear subsection carrying the rear wheel. The rear section bolted to the strut at the rear of the seat-mount casting and to the casting for the lower rear engine mount, known in the factory as the twin tomahawks. The 3 qt. oil tank was on the rear downtubes aft of the rear cylinder head. This is the early version, 1952-53, with short rear frame section and long forks, and it handled it terribly. The pegs bolted to the sprocket cover behind the timing case cover, and the gear lever was on a shaft that ran through the timing cover. Harley-Davidson

Inside, very little interchange. In that first year, the KR used K flywheels but with ball bearings on the mainshafts, not the rollers in the K engine. The left shaft was tapered with a woodruff key for the engine sprocket, while the K had splines. The crankpin was oversized; WR spec to begin with, and later increased in diameter again.

KR barrels were cast with larger ports and valves (part of the Class C system) although of course the ports were polished, shaped and revised by the racing shop at the factory and by the private tuners working with or against the factory.

The valves were tipped toward the bore, convergent, again like the WR engine. But although the K had the same choice of flat or roller tappets, with threaded shafts for clearance and with covers for the valve stems between the barrels and the timing case, the tappets were ground square and the camshaft lobes were angled so the cams, running perfectly true with their gears and with the flywheels, also lined up with the slightly canted valves.

Another chance here for the technically adept to wonder how, with the camshafts straight and their lobes angled, they channeled the stress.

To the side, is how. The K engine had bushings in the gear case cover, *gear case* being Harley's term for the cavity housing the gears, cams, drives to pump, breather and ignition. The KR got bearings, pressed into the cover, and the camshafts were turned down to fit inside them. Every time the cam lobe ran into the tappet and lifted the valve, even so, there was an outer force, a thrust to the outside, putting pressure on these bearings. They worked, but they didn't last so replacing the bearings in the cover was one of the things KR tuners got experienced at.

```
Model ........................... KR/KRTT
Year ................................ 1952
Engine ......... 45° V-twin, iron barrels, side valves
Bore and stroke ................. 2.75x3.1825 in.
Displacement ............... 45 cu. in. (750 cc)
Brake horsepower ..................... 40 (est.)
Transmission ........................ 4 speeds
Wheelbase ............................ 54 in.
Weight ..................... 320-350 lb. (dry)
Wheels ........................... 18/19 in.*
Tires ............................. 4.00/3.50*
Brakes (factory) ............... none; drum f/r
*See text
```

The KR used the 4.5 gal. tank from the street-model K, with KR magneto in front of the forward cylinder, where the K had a generator. The black stamping covered the engine sprocket, the primary chain and the dry clutch. Behind them was the four-speed gearbox, which could be removed from the side in post-1952 engines. The rear pad, by the way, was for when the rider slid back and crouched, with chin on the tank—into the paint, as they say. Harley-Davidson

The K oil pump turned at half engine speed, the KR pump at half that speed. This was a dry sump, with separate tank and with a small crankcase filled with flywheel, and with pressure varying hugely as the pistons moved up and down. Scavenging the crankcases, getting the oil off the flywheels, out of the cases and back to the tank so it could once again be sent to the cylinder walls and bearings and gears proved to be tough. Using a quarter-speed pump, along with a carefully timed breather was how they did it. Or hoped to do it.

Camshaft timing was of course very different from the timing and sequences used on the K. The KR needed no low end power; the K didn't work at high engine speeds. But this time, the factory and top tuners were working together, some of the time at least, and several sets of cams were developed in due course. At first, the lift and timing was fairly calm because nobody yet knew how tight the engine could be turned, or how hard it could be leaned on.

The carburetor, just one, mounted in the vee. On the engine's left was a Linkert, the same basic design as was used on the WR, and a carb that could be tuned by the rider—the high speed jet was adjustable from the saddle. Exhausts were separate, with length determined by where the tuner wanted the power to peak. That is, short pipes for high speed, longer pipes for punch coming off the turns.

Ignition was magneto, with brand varying through the years. Drive was off the front of the gear case, and the mounting was where the generator was on the K engine, again just as the WR had done it from the basic W engine.

The clutch was a development of the K clutch, inside the primary-drive housing. The primary-drive cavity shared its lubrication with the transmission via passages between the two, while the crankcase had its own oil supply. This was good engineering, in that the demands made by an engine—the pressures of bearings and the heat of combustion—are different from the demands made by gears with all those teeth coming together. There are good reasons for having engine oil and gearbox lube; in other words, there are good arguments against the current system of having engine and gearbox share a common supply of oil.

There are equally good reasons for a dry clutch, one that had its friction surfaces interrupted only by air. The key here is friction, which is what keeps the drive plates and the driven plates of a clutch from slipping when they're pressed together by springs. Oil

The KRTT was similar, yet different. Road races and TT races were so alike in those days—being on short, irregular and sometimes paved courses, with turns and hills that become jumps if you went fast enough—that one model ran both events. Same frame, engine and so forth, but the KRTT had K brakes front and rear, and a swing arm and shock absorbers attached to the main frame where the KR had struts. And, the heavier KRTT had a seat with damped posts instead of springs. This example has the large oil tank used in long races.

The KRM had a skid plate that bolted to the lower frame tubes and front mount. Judging by the dings on this much-used and unrestored example, the pipes got banged up anyway. Bill Hoecker

is slippery by nature and design, so if the clutch runs in oil it must have stiffer springs to keep the plates from slipping, and the stiffer the springs the more effort required to compress them. One reason motorcycles had foot clutches for so long was to be sure the springs were strong enough for the engine, and that the rider had enough muscle (legs being stronger than arms) for the springs.

Thus, when Harley went to the hand clutch for the K model, which had a wet primary drive, the engineers chose a dry clutch; five pairs of fiber or steel plates held together by six springs, and to keep it dry they put the plates under a cover on the clutch basket. The releasing disc (what some would call a pressure plate) was eased away from keeping the springs compressed by a series of four, yes four, pushrods, running through the output shaft and moved by a worm and lever.

This is out of sequence here, but this clutch worked on the K and KH and early XL (Sportster), except that oil crept through the gasket and past the cover and made the clutch drag *and* slip. So in 1971 the Sportster got a clutch designed to run in oil, with a stiffer spring and more effort, and a clutch that dragged anyway, followed in 1984 by a diaphragm spring that was better.

But even the new clutch wasn't as nice as that first, dry clutch. And because the principle was sound

The third—but much less successful and therefore less well known—of the K-based production racers was the KRM, designed for western desert racing. It was a KRTT except for roller main bearings, higher siamesed exhaust pipes, a longer rear fender, shorter front fender and a skip plate. Harley-Davidson

and the working experience proved it to be as good as the principle says it should be, the KR clutch was always near perfect. When the designers went to the bigger engines, even when those bigger and better engines grew in power, doubled in power, tripled in power, all they had to do was go to seven pairs of plates and stiffer springs, with a clutch that was virtually foolproof, even now.

This wonderful device delivered power to four speeds, as on the K model, and one more than on the W or WR. This is a labyrinth. There were the standard KR internal ratios, closer than the K ratios. And there were the close ratios for the KR. Then came the C ratios and the D ratios, until by the end of the KR model's run there had been twelve sets of internal ratios.

There were four engine sprockets, with twenty, twenty-five, thirty or thirty-five teeth. There were tranmission sprockets ranging from fifteen to twenty-four teeth, and rear wheel sprockets from thirty-six to fifty teeth. There were so many ratios possible from all this that the factory's tuning book was packed with chart after chart, just telling the final ratios possible with factory parts alone.

At the other extreme, the KRTT first came with the standard forks from the K, although one could tune them with springs and fork oil. The rear shocks were similar, if not identical, to the K shocks. KRTT brakes front and rear were from the K.

Rules changes

In large degree, these machines as sold were only a starting point. The rules in 1952 were beginning to flex just a bit, in that racers could modify the parts and could change the parts if their home factory had filed the proper papers and pictures with the AMA. Harley-Davidson was very good at that, so the parts books were packed with options, as in several sets of handlebars, all those gear ratios, easy change between 18 and 19 in. wheels and tires. There was a kit to convert the KR to the KRTT and back; mostly chang-

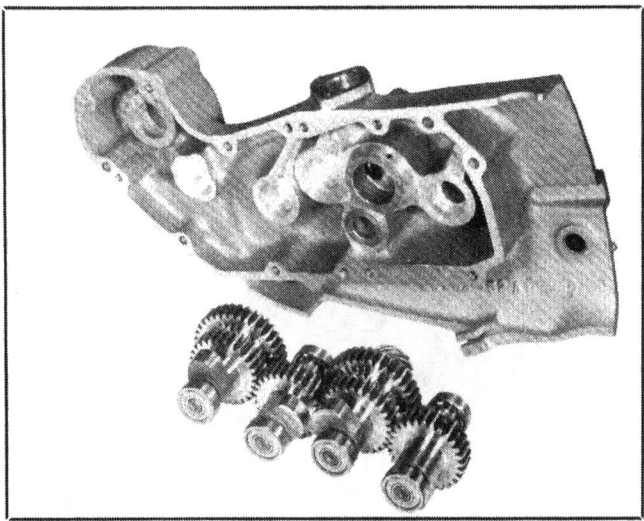

Camshafts were short, with one lobe each. The inner bearings fit into the right crankcase half, the outer bearings into the timing case cover. The cams were driven by the right main shaft. So was the oil pump and magneto. Harley-Davidson

ing the rear subframe or swing arm and shocks, along with the wheels and brakes.

Because the KRTT used the K rear brake, with the brake assembly integral with the sprocket, a KRTT option was a series of rear brake and sprocket assemblies, with a choice of gearing—forty-nine, fifty or fifty-one teeth. And eventually there were selections of large and small gas and oil tanks.

Harley-Davidson's timing was good, in that the KR was right there on stage when the K was shown to

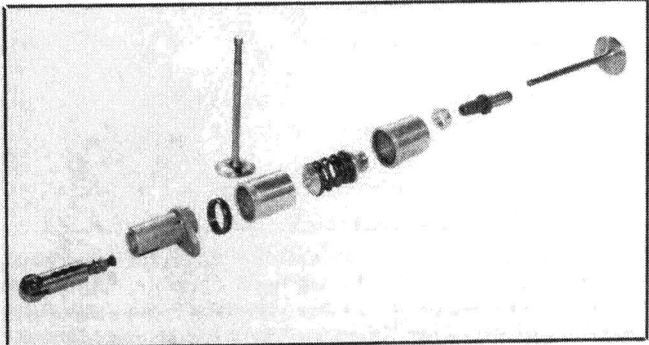

Late KR valvetrain has roller tappet, tappet guide, seal, cover, spring and keeper, cover, keeper, guide and valve itself. Harley-Davidson

KRM, KR, KRTT and K all used cast iron to carry the seat and shock mounts and join the backbone and rear upright frame tubes. The seat was suspended on sliding tubes inside the frame tubes. Bill Hoecker

Flywheels were steel, joined by the crankpin; the pin was kept in place by the jam nut (shown here between the rear rod and the main or drive shaft). The left and right shafts were pressed into the flywheels. This is the left side, where the tapered main shaft ran in a ball bearing in the left crankcase half and carried the engine sprocket, also a tapered interference fit. Harley-Davidson

the public, but the timing was also poor because the KRTT wasn't ready in time for Daytona and the best Harley there in 1952 finished seventeenth.

But, just as the KR began where the WR left off, so did the tuners. Len Andres recalls that in the first year of the KR, the engines came with the same flywheels as on the K. He didn't like that, so he changed over to the smaller, steel flywheels from the 1929 DLD—same setup that worked in the WR. The DLD flywheels were smaller and lighter, so they took up less space in the crankcase and thus created less problem for the oiling system. That made a good half-mile engine, but the light flywheels let the engine come on and off the power so abruptly it played havoc with traction and the rider's concentration, so it wasn't as good on the mile tracks.

The major change or improvement over the long run came because the K had serious problems. The gearbox broke up, and every time the factory had to make the repair, under warranty, the cases had to be split and all the engine parts taken apart and put back together. So the next year, in 1954, the street KH and racing KR and all subsequent Ks and XLs right through to the 1987 model Sportster, got a transmission that came out of the cases intact, through what's

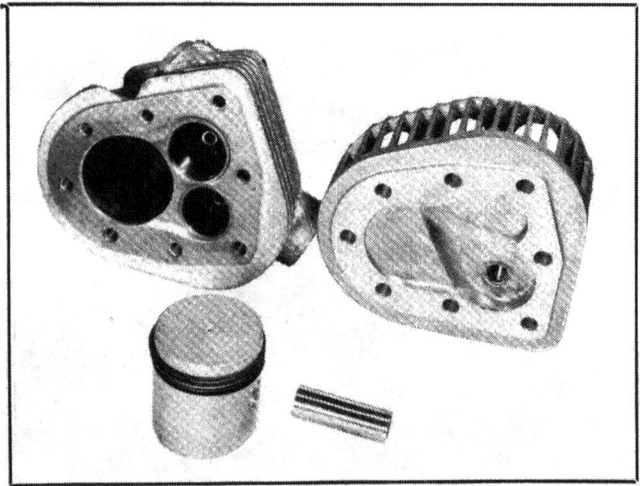

This is the tricky part of the KR engine. The piston, which was flat-topped for all except the 1968-69 road-race engines, wasn't difficult. But the area between the valve seats and the bore of the iron cylinder, top left, and the areas and contours of the aluminum cylinder head, right, were vital. Harley-Davidson

Connecting rods were male and female, or fork and blade. The upper (forward) rod's big end fit into the cavity of the second rod's big end, and the cylinders and pistons were directly fore and aft. The little ends, where the piston pin went, were bushed and the big ends had aluminum carriers with steel rollers. The KR crankpin had tapered ends for an interference fit with the flywheels. Harley-Davidson

now called a trap door, a plate that carries the gearbox shafts and gears and snugs into the cavity at the rear center of the unit. This is handy to have. It's also the way virtually all racing engines are made, so Honda and Yamaha can juggle the internal ratios when needed. And Harley-Davidson offered the feature in 1954, even if it wasn't done for that purpose.

Next came the rules. The substitution rule was obvious. Others weren't; for example allowing riders to modify the tread on tires used for road racing, but not on dirt. Indeed, nobody now remembers how that one came about. Although the bikes came with 18 or 19 in. wheels, the rules allowed the use of any wheel with a diameter no smaller than 16 or larger than 21 in., so the racers went up and down the scale, mostly by running a 21 in. front wheel on the mile. The tall, thin tire had a longer footprint, and the tread offered better grip and the wheel and tire were lighter than the fatter 19 in. combination.

There was freedom to experiment, and people did. Len Andres narrowed the gears in the oil pump so it would move less oil. He also experimented with porting and worked out what was, for him, the optimum compression ratio of 5.95:1. The exactitude of that number hints at just how critical the trade-off was between compression and flow. Dick O'Brien, who later became head of Harley's biggest and best racing effort but at that time was a private tuner and rider, wasn't satisfied with the KRTT in 1953. So his team bought a K and used that frame, stretched and with the steering rake kicked out for stability at speed.

There was no racing team as such when the KR came out. Instead the factory's competition department developed machines for special races: Daytona for road racing, Springfield for the mile and Peoria for TT. These bikes would be loaned out to the top teams for finishing: Syvertson, the department head, liked the Pepper Red color, so all the factory bikes were painted that shade. Len Andres said any bike that wouldn't run was called a "Pepper Red." On the other hand, Brad Andres never lost a race riding on a machine that began life in the racing department.

Most of this is predictable. The KR engine probably had 40 bhp when it began. In 1954 Andres was getting 46 bhp from his best engine, on the factory's dynamometer, while the factory's own engines got 42. There was cooperation, in that the racing shop would call and ask what engine number the team used and then that number would be stamped on the cases sent with the loaner bike.

This kept things legal, outwardly at least, even though Brad used to watch racers stamping cases in the pits and getting caught at it. It got confusing. The Andres team's favorite cases of its own revision were numbers 52KR1498, while the best engine it got from Tom Sifton was nominally 53KR2551. The team had several bikes—for miles, half miles, road race and so on—and the engines moved back and forth from frame to frame and rider to rider. By 1986, when Len

On Daytona's beach, it's Brad Andres (4) leading Joe Leonard (98), both on KRTTs. Daytona Speedway

sold his parts and most of the racing stable left in the family's barn, he still had 1498 and 2551. But while Len thought they still had their original numbers, Brad says they'd been swapped, that is, 1498 used to be 2551 and vice versa, but neither can prove his opinion.

But that's mostly entertainment. In a more serious vein, the early fifties saw some obvious changes that would alter racing, and some changes that would have as much effect, but weren't as visible.

Ready for action. This is Billy Huber, a popular rider in the late forties and early fifties, on a brand-new KR. This was the beginning of another legend: Huber had swapped the 4.5 gal. K tank for a 2.5 gal. tank from the 125 cc Harley two-stroke for use in a short-track race. The cute little tank became a racing item, then a hot street number and finally the only tank offered on a stock Sportster. Harley-Davidson

The obvious began with the dismissal, in 1953, of the proposed Class D rules. Not needed, the AMA decided, because Class C with substitutions would open up the competition and allow other makes (read here, the English) to have machines that would work under US rules. Which they did.

The less obvious involved changes in racing, and in what motorcycle racers or enthusiasts wanted to do. In the West, pros raced on tracks and amateurs raced in the desert, the wide-open spaces. There were point-to-point races across the open land, and there were races on courses rougher than TT, called scrambles. When the course was on unpaved and paved surfaces, the event was called a Grand Prix (talk about getting away from the original meaning).

But that's not important. What mattered was that the events, scrambles or hare and hound, were popular. Some, like the Catalina Grand Prix held on the island of that name, and the Elsinore Grand Prix held near that small town's lake, became legendary.

Further, in parallel with what became the sports car movement, riders of English motorcycles read about and wanted to take part in road races.

Although the AMA sanctioned road races, they were most often smooth scrambles, TTs without a deliberate jump. They were for professionals usually, and they had one class. In Europe, there were races on closed, smooth courses, and there were displacement classes, such as 250, 500 and 650 cc. The AMA was willing to help a bit. In 1953 the club created a

All in the family. Brad Andres, left, William H. Davidson, center, and Joe Leonard. This photo was taken at the finish of the 1955 season. Andres has taken away the national title from Leonard, while Davidson, H-D president, was happy either way. The bike was supposed to have an engine built by Tom Sifton, but it may or may not. More demonstrable here is the 21 in. front wheel the racers used on mile tracks because it had more traction than any of the 19 in. tires available at the time. Harley-Davidson

new class for road racing only, with official sanctions but no prize money, but that was as far as it was willing to go, which will have a marked effect on American racing, years down the road.

The most obvious and most effective outcome was the national championship system. At first, in the pioneer days, the national championships were settled at a championship meet. Then there were selections of national titles, such as Smith is champ for ten miles on a mile track and Jones is champ for ten miles on a half-mile track. Also new was that the winner of the Springfield mile got to wear No. 1 for the next year.

Late in 1953 Don Neal of *Cycle* magazine (at the time the only major motorcycle publication in the US) suggested a points series, like the car people had, and unlike the then-current selection of titles handed out, Neal said, "like trophies at a gypsy tour."

Either his logic was beyond dispute or the fix was in. The 1954 season began with the sanctioning of five road races, one enduro, one TT, six half miles, six miles and one speedway (The speedway was a 100 mile grind at Langhorne, Pennsylvania, on a track that was banked and thus couldn't be a flat track. So the speedway title was for a race nothing like what

Brad Andres breaks the tape at Daytona, 1956. He was top qualifier at 126 mph that year. Note here, aside from the speed and daring, that Andres had his hands on two stubby grips below the normal bars. These clip-ons, as they'd say in road racing, were used at places like Daytona where two long straights were linked by two bumpy U-turns. Brad and Len Andres

Andres on the KR. Notice the small tank and the semi-smooth front tire, which must have worked better on this mile track than the semi-knob did. Harley-Davidson

speedway means to most fans today.). The sanction fee was increased, the added amount put into a fund. Points were awarded for these events on a scale of 5, 3, 2 and 1 and the high-point man at the end of the year would get the extra money, wear the No. 1 plate and reign as national champion.

KRM

Back in the desert, the AMA didn't get involved in what was to become motocross, nor did the national club appeal to riders of smaller bikes. But Harley-Davidson was fully aware of the interest in off-road competition.

This is something of a lost chapter, in that racing fans of a certain age already have heard of the KR and the KRTT, the flat-track and the road-race versions of the competition K model.

What isn't generally known, however, is that there was a third competition K, the KRM, listed in the 1953 brochures right next to the KR and KRTT.

The KRM was mostly a KRTT but with rear suspension and brakes, needed in the desert and in those grand prix races. It came stock with the 4.5 gal. K fuel tank, sprung seat, off-road tires and air cleaner, and with the magneto horizontal on the front of the timing case. For extra protection, there was a skid plate beneath the engine and a guard for the crossover shaft for the rear brake: The pedal was on the left, opposite the shifter, while the brake itself was in unit with the rear sprocket, on the right, hence the vulnerable shaft.

The KRM never quite made it. There were plenty of buyers at first, and Harley ads say that five of the first ten finishers in the Palos Verdes scrambles were

Laconia, New Hampshire, 1958. Brad Andres and Carroll Resweber, on normal KRTTs with 5.5 gal. tanks and Sportster brakes. Notice the semi-knob tires; Laconia wasn't completely paved then. Andres went on to win. Brad and Len Andres

Something completely different: The KHRTT scores its least-likely victory; in 1961 when Fumio Itoh, a member of one of Japan's great racing families, won the All-Japan Motocross. He's shown here vaulting a jump in front of the officials' tent. It was a five-lap race and Itoh won by a margin of a lap and a half! The only visible modification is a tiny front fender. This was the first Harley raced in Japan since before World War II. The record doesn't show if they were ever invited back. Harley-Davidson

KRMs, and that a KRM won the Cactus Hare Chase, but that's as good as the model ever did.

Why and *how* isn't in the books. The magazines of the time were not, how does one say, interested in adversarial journalism. When a KRM came in third behind two Matchless 500s, the reporter said the bike was "highly revamped, greatly improved... the KRM (enduro) was back in fighting trim," but there's no mention of what the problems were. Later *Cycle* titled its test of the KRM "KRM Under Fire," but never said from where the attacks were coming.

The test mentioned that the KRM used flat cam followers rather than the rollers in the KR engine, and that the KRM used more spark advance, a different cam timing, short non-muffling mufflers and a higher (6.5:1) compression ratio than the KR had.

The so-called test had no figures and no dimensions of any kind. Instead, the reports said that the KRM weighed about 50 lb. more than the English 500s, which were 325 lb. or so. The KRM was plainly too much heavier than the opposition, and too big and heavy for that type of racing. The top riders of those desert races weren't interested in riding the KRM, so the model faded away. The KRM was listed through 1955, but few were made or sold and only a handful survive.

Joe Leonard: The first No. 1

Meanwhile, the KR and KRTT went from strength to strength, as in Paul Goldsmith, a factory backed

Carroll Resweber, on the occasion of his winning the Columbus, Ohio, half mile in 1960. Harley-Davidson

Resweber in action. He was fearless and determined and had a style all his own. Harley-Davidson

rider from Michigan, taking Daytona in 1953, the first Harley win there since 1940.

Tom Sifton got a new rider, a Californian named Joe Leonard; a big, strong man capable of winning hillclimbs on his Panhead 74. He rode a KR to third in the Catalina GP in 1953, and *Cycle* commented that Leonard "should go far." And he did, winning the San Mateo (California) mile, the Windber (Pennsylvania) road race, the Sturgis half mile and the Peoria TT (750 class) in 1953.

Leonard won the first series national championship in 1954, with spectacular qualifications. He won eight national races: two road races, three miles, the 750 and the open TT at Peoria on the same day, and a half mile.

The record of eight wins in a season stood until 1986, when Bubba Shobert won nine nationals. The flaw here, for Shobert's case, is that in 1954 there were eighteen nationals, while in 1986 there were twenty-three dirt-track and ten road races. So Shobert had more chances.

But that's merely math. Leonard easily took the title in 1954, with nearly twice as many points as the next man in the series.

Privateers get help

Evolution (later one of Harley's favorite words) continued. The 750 cc K model proved to be not strong enough to keep up with the English machines, which grew into 650s from 500s. So Harley engineers did the logical thing and increased the stroke of the engine, from 3.8125 to a whopping 4.5625 in. This was cheap to do because the flywheels and rods had to be changed but most everything else remained the same, and it increased power where power was useful. Earlier, the factory had released the KK model, a road-equipped K with the KR top end and cams. So with the KH came the KHK, same deal with the larger engine.

For racing, the KR and KRTT were joined by the KHR and KHRTT. This was mostly in the book, as the larger engine was legal only in AMA races for the open TT class. When Harley got a larger engine that was at the same time smaller than the 74s and 80s used until then, the class limit changed from open to 900 cc. The KH was 883 cc and the new limit surely wasn't an accident.

But never mind. The KHR was the KR with the larger engine and the rigid rear section. The KHRTT was the same machine except with the swing arm and shocks and, if the owner wanted, the large or small tank and the various wheels, tires, gears and so on.

The 1955 KR engines had larger crankpins and rollers, because the first versions needed to be rebuilt several times each season. Some of the tuners, inside the factory and out, used to bore out the left case half and install a larger roller bearing. This was a gamble, as the larger bearing would take higher loading, but it wasn't safe to spin it as fast, due to the rotational speed at the outside. But some guys did it and got away with it, while the factory stuck with the smaller bearings all through the KR era. (Cases so modified were called Big Bearing cases. The same thing was done a few times with WR cases, at least two of which survive, but in neither example was such a change recognized by the factory.)

At that time, say the middle and late fifties, the racing shop supervised the assembly of customer KR engines and made sure the engines had at least 42, maybe even 45, bhp when they went out the door. The factory record then was 50.7, remembers former factory tuner Walt Faulk, and when that engine recorded that figure, there was revelry all 'round.

The atmosphere was that of large, temperamental, family. With long memories on all sides.

While the factory was selling customer engines that weren't nearly as strong as its in-house engines, and were clearly behind the top BSA engines (with 45-48 bhp, according to those who should know) the Harley-Davidson booklets were veritable bibles when the true privateer needed help.

In the spirit of Class C rules, the factory issued instruction booklets aimed at the privateer professional racer. The books assumed a certain knowledge of engines and of mechanics, as well as access to professional tools and machine shops. The parts books for the KR (as well as the WR and, later, the XR) listed only racing parts not shared with the road bikes. If the parts were shared (for example, the seal on the left main bearing), the part wasn't in the racing book, on the unspoken assumption that the man reading the book was already familiar with the street machines and/or would know that there was supposed to be a seal on that bearing. (The innocent, the new racer working on his first Harley, was naturally likely not to know about the seal; without one, the engine pumps oil from the crankcase into the primary case and the gearbox—about one quart every fifty miles. I speak from experience.)

But that's only part of the story. The best part, the important part, was that the instruction booklets were precise where it counted. The actual bore size for a new KR, the book says, wasn't the 2.75 in. listed in the brochure. The actual, carefully selected bore size was 2.747 in. The KR cylinders could safely be bored an additional 0.045 in., to 2.792 in. and still be within Class C regulations. The boring was done in stages, first over, second over and so on, just as with any other motorcycle of the day.

The booklets went even further. Racing engines have always been special, but these went well beyond routine racing practice or the hop-up techniques of the day.

First, the factory said, the new engine should be run for about fifty miles, with revs under 6000. At the end of the run the timing, valve clearances, all torque settings, nuts, bolts and so forth should be carefully checked and corrected if needed. Then the engine could be raced. Oddly here, no redline was listed or

Seeming a bit apprehensive, Resweber with an early XLR, 1959. A master of the half mile, at home on a road course, Resweber somehow never won a national TT, which is what the XLR was built for. Harley-Davidson

suggested, although usually the tuners kept below 6400 until the engine's very last years.

With 100 or 150 racing miles on the engine, it was time to take off the cylinders and have them bored (not honed) to the first oversize, and time for the valves and seats to be refaced.

But wear wasn't the problem here. Instead, the factory tuners had learned by doing that the cases would shift and the barrels would distort, get out of true just the smallest bit, during those first few miles of operation. So the engine needed to be taken down, corrected and put back together.

Harley's booklets provided all the clearances and settings for the Linkert carburetor that was fitted on the first KR, and for the Tillotson that came in 1966, as well as for the dual manifold, carbs and linkage, right down to the name of the outfit that made the ball joints in the throttle cable block when dual carbs appeared on factory bikes.

Probably the most important part was the set of sketches with which a skilled, careful machinist could recontour the ports, valve seats and the cylinder relief area (the area around the valves, between the valves and the bore). There were sketches for the ports, and for the cylinder heads; during the KR's production life, 1952 until 1969, there were running changes to the crankpins, cases, barrels and heads, of which seven different designs were made. They'd all backfit, of course, but even with the seventh set, the factory offered blueprints on how to make them just that little bit better.

That was both the point and the bottom line. The side-valve engine (the principle instead of just this application) was a sound choice in the thirties, reasonable in the forties, outmoded in the fifties and the result of stubborn miscalculation and politics in the sixties . . . and Harley racing used the design until 1970.

The company did it against the odds, against the trend. The only reason it worked is that the men who made it work, put so much into it. And they were smart. And there were a lot of them and to some extent there were a lot of them because of the factory booklets.

The man who bought the KR engine and raced it got at least 42 bhp. The man who bought the engine

The XLRTT was an odd combination. The frame, brakes, suspension and so on were from the KRTT but the engine was based on the XL, the Sportster. The XLR was the same displacement as the KHR, but had different bore and stroke, along with overhead valves. Inside, the XLR used ball bearings, hot cams, magneto ignition—all the racing stuff. Harley-Davidson

and the booklet for that engine and followed all the instructions got the 50 bhp the factory had. The man who read the book and listened to his brother-in-law or the guy across the street, blew up the engine. And the tuner who read the booklet and thought about it and did his own thinking... and won, got help from the factory, even when there was rivalry.

Brad Andres: More than his father's son

Harley-Davidson's second Grand National champion arrived as almost a sports cliche. Brad Andres was a member of a motorcycling family, son of a man who owned a collection of Harley dealerships. Len Andres was a good racer—not the best, but good. And so were his brothers. So, like Mike Hailwood, Stirling Moss, Mickey Mantle and Bob Feller, Brad Andres got into a familiar sport, and began with either crushing pressure or an inherited advantage, or both.

Circumstances also came into play. American road racing wasn't like European road racing; import fans formed their own clubs and put on races with classes for displacement and with no regard for Class C rules. The AMA opened the door a crack and sanctioned championship road races, for Class C bikes, of course, but on real road courses, such as Willow Springs, California.

"American road racing was reborn on April 4, 1954, at Willow Springs" said *Cycle*. The winner rode a BSA Shooting Star, a twin. The best Harley came in eighth, but in the Amateur race, Brad Andres led until his KRTT broke down.

Brad Andres may have inherited riding skill. More likely he soaked up his father's interest and brought his own drive to the task. Andres was high-point AMA Novice in 1953, the top Amateur in 1954 and made Expert in 1955, whereupon he won Daytona, then Laconia (still run on the old, TT-style course), then Dodge City, Kansas. He won the mile on the old Langhorne speedway, finished the year in his backyard, winning at Torrey Pines, just up the hill from the Andres' real estate conglomerate in San Diego. Brad was No. 1 for 1955, well ahead of Everett Brashear (also factory backed) and Joe Leonard.

Langhorne was banked, so it was declared a speedway rather than a flat track. It was irregular, sort of like a D, and it was uphill and down. The track surface broke up during the race and riders were

Resweber and Bart Markel, who played follow the leader into the national title. Visible here are the steel shoes, actually skid plates that strapped to the left boot, used for balance on the left-turn-only mile and half mile. Also, the KR's carb and massive air cleaner stuck out a long way. When the bike lowsided, slid out from under the rider, the carb was the second thing to fall off. Harley-Davidson

bombarded with chunks of flying mud, rocks and other debris. Most of the racers hated the place. The AMA Competition Committee finally had an excuse to yank the track's sanction and national races were never held there again.

But Andres was a tough cookie. He loved Langhorne: "Most of the track you could run wide open." Andres wasn't a showman though. He didn't appeal to the crowd, he wasn't spectacular and he didn't have much of a grin. An intensely private man, he was also intensely competitive. Some might have raced to ride, but Andres rode to win.

He was a product of his time, in that the Harley-Davidson instruction booklets focused on the engine. There was nothing about chassis tuning, so the riders with some knowledge or appreciation of handling could use it to advantage, and Len and Brad, father and son, had literally generations of accumulated skill.

There was another family aspect. During this period, the fifties in general, was Harley-Davidson very much on top. In 1954, first year of the points race, there were eighteen nationals and Harley won thirteen, to five for BSA. In 1955, there were seventeen nationals and the score was Harley fifteen, BSA and Triumph one each. For 1956 there were seven nationals, H-D won them all; in 1957 it was eight, with six to Harley and two to BSA.

This wasn't quite the total lock it appears to be in the books, but not by much. The Andres, Sifton and factory racing groups worked closely together, with the factory loaning engines and parts, and the tuners sharing what they'd learned and swapping and buying from each other.

Professional racing was *almost* a good way to make a living. Brad Andres had help, obviously. He also did well with the prize money, and there were some unexpected benefits. While the promoters didn't pay start money, various dealer organizations sometimes did, so they'd know their customers could expect to see the best man riding for their brand. At the end of his championship year, Andres banked $20,000, at a time when that sum was heaps more than it is now, thirty inflationary years later.

Rule refinement

The inequalities hinted at by the lopsided results of KRs versus the others didn't go unnoticed, nor were they uncorrected.

But it was a struggle. The rules were made by the AMA's Competition Committee, which was appointed by AMA officials, who had been hired by the trade group (read here, the American factories) that inherited the AMA in the twenties. Not even members of the committee, according to Len Andres, knew why they'd been picked when they were asked to join.

There was a rough sort of balance. When there was Harley-Davidson and Indian, each factory got to have the same number of members on the committee. When there was just H-D and the importers, the other brands got a sort of parity. In 1953 the committee was composed of thirteen Harley adherents and thirteen independents. In 1954 the split was fourteen to twelve, and in 1955 it was fifteen to eleven. This isn't to say all members were strictly partisan. As a group, they were interested in the sport and surely voted their consciences. (It's also worth noting that people generally manage to convince themselves that what's good for them is good for the world.)

Late in 1955, a rider's committee cornered members of the AMA committee and submitted several demands. First, the Competition Committee should be evenly divided, thirteen for Harley and thirteen for BSA, Triumph and Indian (still in business as an importer of English bikes). The deciding vote in case of a tie could come from the AMA's executive director.

Markel and wife Joanne after a half-mile win, 1963.

Second, the limit on the compression ratio should be raised from 8:1 to 9.5:1; and, third, the purses should be bigger.

The AMA officially took no note of the first, and the third was endorsed even though nothing was actually done; in fact would never be done except that much later R. J. Reynolds Tobacco Company began putting money into racing.

The second issue, the compression ratio, ostensibly was there so real production motorcycles could compete with one another and have a chance against specialized racing machines. If it happened that overhead valve engines took over the rest of the world so that production machines were running much higher compression ratios than before, and if it happened that side-valve engines didn't gain as much from high compression ratios and so lost less from the limit, well, that's the way it goes. Or went.

This sounds impossible today, but in that 1956 meeting a formal report was requested by committee member William J. Harley and presented by his son, William S. Harley. (Yes, of course, it was *that* Harley family.)

William S. Harley said the Harley-Davidson engineering department had obtained a 500 cc ohv racing engine with the then-prescribed compression ratio of 8:1. The engine delivered 40 bhp on the factory's dyno. The engineers then changed the pistons and raised the compression ratio to 9.5, which boosted output to 42.5 bhp.

Harley (the man, not the factory) said that a 1955 KR engine as delivered to a customer, cranked out an average of 42.5 bhp. A complete KR weighed between 342 and 350 lb., depending on tires, brakes and equipment, Harley said, so with minimum weight and average power, the KR had 8.04 lb. per bhp.

A Triumph T100R, with rigid frame as used for Class C track racing, weighed 307 lb. With 40 bhp, this gave 7.66 lb. per bhp. A race-prepared BSA weighed 285 to 295 lb., so it had 7.12 lb. per bhp.

Quoting Harley's report, "It can be seen that the 750 cc side-valve motorcycle is actually at a disadvantage."

Perhaps. Left unmentioned was torque and midrange power, the punch off the corners that meant as much as the top-end power delivered down the straights. Nor (as noted) were the top Harley engines limited to 42.5 bhp. If the factory's best had 50 bhp and Andres and Sifton machines weighed closer to 320 lb. (or even less, but probably not quite yet) then the KR's power-to-weight was 6.4 lb. per bhp, the sort of ratio that could help a chap win all the nationals, which is just what Harley-Davidson did.

There was something else, though, worth hearing now that the sounds of this battle have faded away.

Harley's report reminded the committee that Harley-Davidson didn't have the upper hand when Class C began (true, as we've seen). "During this period we at no time requested rule changes to alleviate this situation. We continued to work on our equipment and after a period of some years we began to win some nationals . . . in the past couple of years we have been successful in our efforts and feel that our competition should be willing to spend the money and effort to improve their equipment and provide riders and tuners . . . we feel that the rules as established are equitable and just."

The equity here could still be debated. But it does reveal the feelings behind the history behind the defense.

And the minutes show the system to have been more equitable than supposed.

There was a motion to allow 650 cc engines with compression ratio limited to 8.5, defeated 16 to 11.

There was another motion to allow the 500s unlimited compression ratio, defeated 16 to 11.

Quoting from the minutes, "At this point (name deleted in the published account) asked to be relieved of his duties as a committee member."

Then there was a motion to have the 500s race in a separate class, defeated 17 to 9.

There was a motion to allow the 500s a compression ratio of 9:1, approved 15 to 12.

That's how the system worked, with some committee members on this side, others on that side, the guys in the middle voting their consciences often enough to have things balance out. The importers took heart, they began offering models for flat track; rigid frames (approved from above), the whole thing. The imports won no nationals in 1956, two in 1957, four out of ten in 1958 and four of ten in 1959.

Back to normal business, the committee voted to limit the nationals to one type of each event; for example, one national on a half mile, one on a mile, one TT, one national enduro, one "Class A" hillclimb and one road race.

This was odd. The road-race national went to Laconia, New Hampshire, provided (no kidding here) that the promoter graded portions of the course and removed two trees the committee felt were hazardous.

Daytona Beach didn't get a national sanction. That race was held, on schedule, but it was billed at the time as a Classic, a vague term now in vogue for golf and tennis shows. The 1956 Daytona Beach race was not, according to the press reports and AMA bulletins, a national championship race. It was run, and won by John Gibson on a KRTT. The AMA record books show it as Gibson's only national win, and show Daytona in the national points compilations.

Len Andres, who was on the committee, thinks the problem was that the promoter couldn't come up with the cash in advance. Or, maybe it was a clash of personalities. E. C. Smith (AMA executive secretary) was a man who wasn't afraid to insist on his own way. However, Andres says, starter Jim Davis either reached into his own pocket or used his excellent reputation, because the money was posted and the race run.

What really was important, Len says, was that his best ever KR engine, ol' 1498, was tops in the field. Brad was clocked at 126 and got the pole. It was the best engine, a happy engine, Brad says, too bad the mag drive wasn't right yet and came loose and Joe Leonard's cam train went out and Gibson, backed by a good dealer but not by the factory, won.

To finish the committee meeting in the traditional way, the members voted not to sanction Langhorne for a national, because the increased compression ratio allowed the ohv 500s would give more power and more speed than the track was built for. The vote on that one was 23 to 2, so it wasn't imports versus domestics or Harley against Them, but a vote on a grudge.

Consequences

That the 500s didn't do better than they did with the gain in compression and in influence and with practice, was due, in part at least, to two factors beyond their control.

One was the retirement of Hank Syvertson ("racing department" coordinator) and the arrival of Dick O'Brien. It was in March 1957, just in time for Daytona.

Syvertson was an able, amiable man, who was diligent but overshadowed by Walter Davidson, a racing fan and a man willing to overrule his employees when they conflicted with his strongly held opinions.

O'Brien was able and agressive. His title was racing engineer, although his engineering training came from the working end of a wrench, rather than in a classroom. He'd been a poor kid who made tools and traded them to the local motorcycle store for time on bikes. He raced sprint cars and motorcycles well enough, but was never championship material. He was ninth at Daytona once, and blew up there once. Like most great tuners he came to grips with his limits and worked from his strengths, in this case tuning for a big dealership in Florida.

O'Brien's specialty, of all things, was two-strokes. He was tuning winning Hummers and 165s before the factory even paid attention to them. O'Brien had lots of ideas and was willing to express them and he proved good at spotting talent. He was rough and tough and he was also warm and outgoing—leadership material. And yes, did he ever make enemies!

The late fifties were something of a technical plateau. There were new cams for the KR, and a better valvetrain and new heads. But the gains were measured in small increments and details, like fitting second sets of grips, clip-ons they'd be called later, to the fork tubes, below the bars, so the riders could tuck down on the tank and pull their elbows in on the straights. The K model brakes got scoops. BSA and Triumph listed dirt-track models, sold without brakes and with rigid frames.

The AMA announced details of the Sportsman class: full road equipment, at least fifty examples produced, ridden by the owner and registered in the owner's state, and approved tires only, no brakes in dirt events ... and displacement classes with 175, 350, 500, 750 and open displacement. No mention of valve gear, but with a 9:1 compression ratio limit.

The jostling was mostly at the top. Joe Leonard won the national title in 1954, Brad Andres in 1955, Leonard again in 1956 and 1957. Everybody swapped back and forth. When Brad was clocked at 126 on the beach's short straight, Joe turned 124 mph and the best Triumph did 121. The legendary Jackpine Enduro went to a lightweight, a highly modified Harley 165. Andres temporarily retired with a badly injured leg, about the same time Tom Sifton got tired of hearing that he won only because he tuned Harleys. Sifton went into the BSA business and rider Ev Brashear went with him, putting BSA into the win column, while Leonard rode for Len Andres at Daytona and won the 1957 race, which was back as a national.

Carroll Resweber: Four in a row

Two chance meetings: In about 1952, Carroll Resweber, a kid from Port Authur, Texas, went over to Houston for the races. He rode there on his Knucklehead Harley, stroked to 80 cu. in., with flywheels from the flathead UL Harley. He knew motorcycles, but he'd never seen a race.

He walked over to the fence and leaned in, just in time to hear a terrible noise and see an apparition, a cloud of dust and smoke coming straight at him.

"I ran for my life," Resweber says now. When he'd picked himself up, he looked back and saw Joe Leonard headed into the second turn, still sliding at 100 mph.

He's never forgotten his introduction to racing. First, he says, it was a powerful, emotional experience when he made Expert (at the end of 1956) and was out there on the track, racing with and beating his heros from that day in Houston.

For the rest of his career, every time he saw a kid leaning on the fence, he'd run into that corner as hard and deep and close as he could, to make sure other kids got the same thrill that had meant so much to him.

The second chance meeting took some time. Resweber borrowed bikes and worked his way into his professional license. His talent showed. He got Brashear for a coach, and was set to use Paul Goldsmith's machine, as Goldsmith was getting ready to move into stock car racing, but Goldsmith wrecked it before the transfer.

Resweber went to see a man named Ralph Berndt, who worked for Harley-Davidson but raced against the factory at the same time.

Berndt was good enough to have been made in Hollywood. A taciturn, wryly entertaining man, he was mostly a machinist and a theoretician—a careful worker. Berndt liked fights and didn't suffer fools for long.

Once again, Berndt had been a rider "until I went on my head once too often," he says. He'd ridden an Indian Scout, so he switched to tuning Scouts, even though he worked for Harley in the frame shop of the assembly plant in Milwaukee.

Syvertson and Walter Davidson noticed Berndt at the races and persuaded him to transfer into the racing shop, and to begin tuning Harleys, the other way being something of an embarrassment.

He remained an embarrassment. Berndt worked days in the racing shop, doing what they told him to do, then worked nights on his own, usually with rejected parts and ideas offered to management but turned down. "I was kind of a thorn in their sides," he says now.

One of the leftovers was a set of engine cases, stamped 52KKX8. This, in Harley-Davidson code, means an engine made in 1952, a tuned version of the road-going K instead of the racing KR, with X standing for experimental. Berndt bought the cases, switched to the ball-bearing mains, big crankpin, quarter-speed oil pump and the rest of the racing stuff; installed his own cams; and built his engine his way. The engine was in Berndt's shop when Resweber walked in looking for a ride.

The rest, as they say, is history.

Resweber won the national championship in 1958, 1959, 1960 and 1961, a string nobody has equaled since. He didn't walk away with the titles, there were plenty of close finishes. Resweber won the title over Leonard by one point in 1958, and four points in 1960, and beat BSA team rider Dick Mann by fifteen points in 1959. Resweber won miles, half miles, short track and road races—everything except TT. No special reason, he says, he just didn't do it.

Racers from different eras can't be equally compared, as there are too many variations in rules, the strength of the competition and so on. But the best from several eras have some things in common.

Style, for one. Those who saw Resweber in action never forgot it and never mistook him for anybody else. Dick O'Brien says Resweber wasn't so much a natural as a determined man who taught himself how to win because he wanted so much to win.

Case in point: Two nights before a major national, a road race, Resweber and the guys from the local Harley store, where they'd been preparing the bikes, went out on the town. Resweber was on a borrowed machine, a BSA. He had had too much to drink and forgot where the road went and crashed. His collarbone snapped. As was common in those days, he didn't tell anybody. Resweber walked the course on Saturday, then qualified thirteenth. Not bad considering he couldn't push or pull with his left arm.

He fell after the start. On the way down he thought, "Well, I'm glad this day is over."

But it wasn't. A fan rushed over, righted the bike, kicked it straight and got the engine running, so there was nothing Resweber could do except get back in the race. "I want to get off this thing," he said to himself, so he began shifting on the kill button, throttle wide open and clutch lever untouched.

The KR was a funny engine. Once Andres had a dragging clutch, so he tried to pop the gearbox into first and the whole thing came apart. This time, engine wide open, tach needle on the peg, "The thing went the whole 100 miles. I finished ninth."

His epic ride inspired half the crowd, or so it seems, to slap Resweber on the collarbone. The broken one. He went to the hospital to have it pinned, then drove himself home. He rode in a half mile the next week, finished second. When the pin was removed, it had been bent by the leverage Resweber exerted on the bars.

"You want to race," he says, "pain means nothing."

Along with being fast and tough, Resweber was personable. He adopted the small town of Cedarburg, Wisconsin, and the town adopted him, with a banquet and the gift of a new car when he won the title.

Resweber was also crafty, and no mean mechanic in his own right. He did the chassis and Berndt did the engine. They worked on their team at night, as there was no Harley team. Walter Davidson oversaw the racing shop, with Syvertson and, later, O'Brien.

From the outside, it sounds like an extended, combative family. William Davidson was Harley-Davidson president, but as Brad Andres says, "Walter *was* Harley-Davidson . . . If you wanted to talk, you talked with Walter."

Or you listened to him. Berndt worked in the racing shop but not on his machine. Resweber worked outside during the day but did get paid by Harley-Davidson, for expenses and a bit extra now and then.

There were no team colors, except that Syvertson liked red. The Ohio riders, led by George Roeder, liked yellow. In fact, some racers painted their bikes yellow so people would think they were from Ohio and therefore fast. Berndt and Resweber painted their best machine blue. One of the guards at the Harley plant was an amateur wildlife painter, ducks and deer and such, the art one now sees in decals on the backs of camper shells on trucks. He painted a goose in flight on the tank of 52KKX8, which was done in blue and white and which was known, naturally, as "The Blue Goose."

That was all the bike had on it. Resweber grins that one day Walter Davidson came up, walked around and around the Goose and asked, deadpan, "What kind of motorcycle is this?"

"I was on the payroll," he says, "and I didn't need any further hints. The tank came off."

He and Berndt kept their advantage in other ways. They thought about things a lot. And Berndt was the sort of man who liked trips to California because he could go to the Douglas Aircraft backyard, where they threw away chunks of aluminum and didn't mind if people picked them up and took them home.

Where the other teams used steel, Berndt had aluminum. Other racers had rubber grips on their bars, Resweber had strips of tape. It sounds almost silly, except that they got their KR down to 261 lb.

It all sounds logical. They heated and bent the KR's rear subframe so it would be lower and have more weight in back, just the way the second man carrying a couch up stairs finds the load has shifted to him. Resweber experimented with a disc rear brake on his KRTT in 1961, years before the factory did it. But he still believes brakes, shifting and suspension chopped up the track, narrowed the groove and therefore damaged the show.

What about stopping?

"You ever rollerskated?" he asked.

"Uh, sure." I replied.

"Same thing. Cock it sideways and slide to a stop."

Which of course is easier to say than do, except that Resweber used to throw the rear wheel out of line to slow down, full sideways to stop, or he'd cock the front wheel into the turn to slow and then hang the back out with power . . . and he made it look easy.

Resweber also worked hard. During 1962, inbetween the national races, which took his weekends, and after regular work as a welder, he raced short track in the Milwaukee area. He made eighteen races, won sixteen and collected $106 every Wednesday night. Prize money from nationals totaled $18,000 in 1962, with the rider getting sixty percent, the tuner forty percent. In short, they needed their regular jobs.

Then came the crash, on September 16, 1962, on the half-mile track at Lincoln, Illinois. Five riders went down during practice on a dusty track. One man died on the way to the hospital, another retired from racing that day.

Resweber doesn't remember a thing: "The next thing I knew, I woke up in the hospital a month later."

Markel on the mile, 1966. He was national champ three times, a tough, determined man. As O'Brien says, "You had to kill him to change his mind." But most of the time Markel was right. Harley-Davidson

When the doctors allowed it, Walter Davidson flew Resweber home. He was in a body cast, didn't walk for two years and had lost the use of his left arm. He regained the use of the arm, although the muscle never returned.

Resweber never raced again. He went out and practiced, but the natural moves, the trained reflexes, were gone for good. He was hired for the racing shop as a fabricator, and was the main constructor of the team's Battle of the Twins racer (described later), then was laid off and went back to welding in H-D's economy drive of 1986.

Carroll Resweber has one regret, and it's none of the things one would expect. He doesn't mind not winning all of the five types of Grand National race, or never winning Daytona, or even not being the first and only man to win five straight national titles, although he was well in the lead on points when he crashed.

Resweber still would like to have won the Springfield Mile in 1962. It would have been his perfect race, as well as clinching his fifth No. 1 plate.

He was hot. "I couldn't do anything wrong. I couldn't make a mistake," he says.

Or so it seemed. Clyde Denzer, second in command at the racing shop, says it was typical Resweber. He took an immediate lead, and you could hear that distinctive engine howling out of sight down the far straight. He caught up with the stragglers, then lapped the field. He looked up and there was No. 8, Bart Markel, the second-place man drafting back toward him.

Then his engine coughed. He reset the main jet, which you could do from the saddle then, but the engine coughed again and died. No gas. Somehow one of the rubber pads between tank and frame had been left out, the tank had rubbed through against the frame and enough gas had seeped through to empty the tank. That was that.

After the crash, Berndt had some other riders but it must not have been the same. He put 52KKX8 in the corner of the shop, in case he ever wants to restore it, and went into working on experimental aircraft.

"It wasn't hard beating the factory," Berndt says. "They built like they were going to war."

The competition gets tougher

Resweber's dominance of the top tended to mislead, in that everybody, on Harleys or other makes, backed by the factories or not, was learning and going faster.

Some of this was due to production bikes, just as the Class C spirit would have it. When the stroker KH created the KHR, with more power and thus more speed, the KR seemed to be unstable at top speed. So the factory introduced a second, longer subframe, with a nominal wheelbase of 57 rather than 54 in. (Other factors being equal, a longer wheelbase is more stable at speed, while a shorter one steers quicker and goes around an arc, or corner, with less lean per degree of turn, therefore you can corner faster.) The Andres team used this choice of subframe and wheelbase, and a selection of fork clamps, called triple trees in the trade, and fork tube lengths and offsets between steering head and clamps to tune its chassis.

Next came lightweight frames, made of 0.040 in. wall tubing instead of the 0.060 that the first versions used. There were long and short fork tubes, steel and aluminum fork clamps. By 1960 the KRTT had an option of larger brakes, and the front brake could be adapted to the rear wheel, on the other side of the hub rather than tucked inside the sprocket housing.

The first result of this was closer racing. Where Leonard once lapped the field twice, on what *Cycle* said was "The best K model ever produced," and Andres lapped the field at Langhorne once, Resweber *almost* lapped the field at Springfield.

Second, the imports were catching on. And getting more of a fair shake at the same time, as the racing Nortons finally were approved for Class C and the rigid BSAs and Triumphs came on the market. O'Brien says the removal of the compression ratio rule was a gain, mostly for the British tuners but also for the rest of the circus because there was no more time wasted checking compression ratio.

Third, it drove the AMA officials nuts. All those parts, where it used to be only what came for the street from the factory. Earl Flanders, who was the referee and thus responsible for inspection and enforcement, used to mix nail polish colors, then dab the mix on the cases and elsewhere. Heat from the engine would change the color and it could never quite be duplicated, so the parts couldn't be swapped illegally.

The short-rod theory

Along with the minor tricks and variations came some hard, deep thinking. The side-valve tuners always knew they were at a disadvantage. O'Brien is the first to say that if BSA, Triumph and Norton had worked as hard or spent as much, they would have run Harley-Davidson off the tracks.

Instead, it was Sifton and Andres and O'Brien and Berndt and Lawwill and Axtell and Branch who read books and stayed up late in the lab.

Along with working out camshaft profiles and sequences, ramp speeds, port shapes and contours, came things like short-rod engines.

This is easier to believe in than to explain. To begin with high school physics, the crankshaft, connecting rod, piston and cylinder are a way to convert up-and-down motion into rotary motion: The piston goes up and down, the crankshaft spins.

Next, imagine a circle. Inside that circle, is a square, perfectly parallel to the ground. In the middle of the circle and square is a dot, perfectly centered. The dot is equidistant from every point on the circle and from the four points where the circle and the square touch each other. If you went around the circle, you'd find that the four points are ninety degrees apart, again equidistant from each other. If you travel around the circle at a constant speed, it takes the same amount of time to go between any of these two points.

Fine. But now draw a point above the circle and square, say three or four times as far from the circle as the circle is high.

At wide-open throttle, picture a connection, call it a connecting rod, with one end moving up and down at this point above the circle. The other end—you guessed it!—is going around the circle. It's just as far, going around, from the lower left point to the upper left point as from upper left to upper right.

But, the catch is, when our connection goes from lower left to upper left, it goes up a lot farther than it goes up and down from left to right.

What's all this mean? Mostly that while the crankshaft runs around at a steady rate, the piston doesn't go up and down at the same constant; it stops at top center and bottom center. It slams to a stop, sits there and is yanked back to speed.

Back with the squared circle, the crankpin spends twenty-five percent of its time going across the top, but the piston doesn't travel twenty-five percent of its

Subtle stuff here. Look very closely at this KR engine. What's missing? The housing for the magneto drive at the front of the timing case; it's been trimmed off. Why? To provide clearance for the front exhaust pipe, which is lower because the barrels are shorter, and the pistons are closer to the flywheels because the connecting rods have been shortened about an inch. This is an extreme example of the fabled and perhaps useful short-rod engine. (This example was built on the West Coast, loaned to the factory shop, ridden to a top time by Roger Reiman, returned to the builder and restored to show condition.)

up-and-down motion, its stroke, when the crankpin goes (1) around, from the crankshaft's point of view, and (2) across, from the top of the cylinder's point of view.

Got that? Next step is that all this varies with the distance between the center of the circle and the point of reference, that is, with the length of the connecting rod. And there are variations in angularity. Picture the piston in the cylinder above the flywheels. With the crankpin at midstroke, full right, if the rod is short it's cocked at an extreme angle between piston and crankpin, If it's a long rod, the angle is less. While at top and bottom centers, there's obviously no angle, no matter what the length of the rod.

Adding to that, because the angle when the crankpin is nearly at the top isn't the same as the angle when it's nearly at the bottom, the bottom half of the stroke, so to speak, isn't the same as the top half. In terms of time, the piston is traveling at its top speed approximately 76 deg. before and 76 deg. after top dead center (TDC). (This also varies with the length of the rod.) Because of that, the piston goes from maximum speed to zero and back to maximum in the top 152 deg. of crank rotation, and uses 208 deg. in the lower half of the same sequence.

Furthermore, the speed variations and time spent at various positions influence the cam and valve sequence. The best opening and closing times for an engine with a given connecting rod length won't work as well if the rod is changed because the piston won't be traveling the same speed or in the same place at the same crankshaft position.

Fascinating, you say. And so what?

Several of the top tuners worked this out before it appeared in books. They decided that shorter connecting rods would give more leverage on the crankpin, and therefore more power, and that the shorter rods would keep the piston doing useful work longer. The length of time when the burning fuel and air are expanding and forcing the piston down and the crankshaft around can vary with—guessed right again—the length of the rod.

Sounds simple. The tuner cuts the stock connecting rods in the middle, at an angle, removes an inch or half an inch, whatever he thinks he can get away with, and welds the rods back up.

Not so fast and easy, there. That is how the rods are shortened, yes. And if they were welded skillfully and carefully, they will work. It's obviously better if the engine builder can find a fabricator or machine shop or specialist—Carrillo, for instance—where connecting rods are made for racing from scratch, but the stock parts can be reworked.

In the case of the KR, they had to be reworked because racing parts weren't legal. When short rods first became known, the guys who weren't building them raised objections against those who were and, the AMA Technical Committee had to consider the

Substitution time. This is an early all-tube frame, photographed in 1962 for recognition and approval by the AMA for the 1963 season. This frame has the normal twin tubes around the engine and shows the twin tomahawk, the paired castings that contain the swing arm pivot and the rear engine mount flanges and ledge, and the lugs facing the rear, used to attach whatever shows up there. But the steel steering head, with brace, the fabricated tubing for the backbone and rear tube junction, and the seat and shocks are new. Note the angle of the backbone and the location of the head steady, the brackets below the center of the backbone. Harley-Davidson

Rear disc brake was more than experimental; it had steel calipers with the lining material on the disc instead of on the pucks. This brake was submitted for approval in 1961 but not used in any races. (Several racers tried their own versions of disc brakes for the KRTT.) Harley-Davidson

question. The short rods were legal, the vote went, because they didn't change the displacement and because the parts were reworked production parts.

A lot reworked, in fact. The shorter rods meant the pistons didn't go all the way up the cylinders, so the barrels had to be cut down. If the engine was a vee, which the KR was, then the intake manifold had to be trimmed because the shorter barrels put the ports closer together. The exhaust pipes had to be revised, as did the head steady between the cylinder heads and the frame. If you went short enough, the front exhaust pipe ran into the timing case cover, so that had to be trimmed back, and that meant the magneto had to go atop the case rather than in front.

What all this means is a lot of work.

In theory, shorter connecting rods also mean more power. The shorter the rod, the less time the piston spends sitting still and the more time it spends going up and down, and the sooner the crankpin is at a useful angle to the piston, creating more torque.

Working against this, so to speak, aside from the bother of making a rod shorter instead of beginning at the design stage with shorter rods, is the obvious handicap of increased angularity: If the piston pushes the crankpin sideways, the angle of thrust in turn jams the piston sideways against the cylinder wall. More stress, more wear, shorter engine life.

Pause here for breath. The tuners did make short-rod engines. They did work, to some degree. The engine shown here was built on the West Coast, loaned to the factory for observation and used by Roger Reiman to set top qualifying time in at least one race. And it's still running.

Len Andres and Bart Markel, among several, made their own versions of this project. Brad Andres has vivid memories of his dad's short-rod engine, and introduces a factor the books don't talk about.

Remember the oil system, and how the pistons going up and down created alternating pressure and vacuum in the crankcases? If you shorten the rods and cylinders, if you reduce the space below the pistons while the space above the pistons stays the same, you get more variation in pressure.

Brad Andres says it was really something. The short-rod engine had more power. He'd thunder past

George Roeder, whose peers said he was the most stylish slider of his day. Harley-Davidson

the other guys in practice, then the crankcase would get too much oil and the flywheels would bog down, and the other guys would thunder past him... The oil would clear out, get pumped into the tank where it belonged, and he'd pass them... the cases would fill again... and so on. Not fun and hard to explain, he says. That experimental engine cost Andres several wins while, although it did make extra power, say two or three horses, it never won any races. On balance, in retrospect, Andres wishes his dad had stuck to what he knew worked.

The short-rod concept, though it never quite paid off during the KR days, will crop up again.

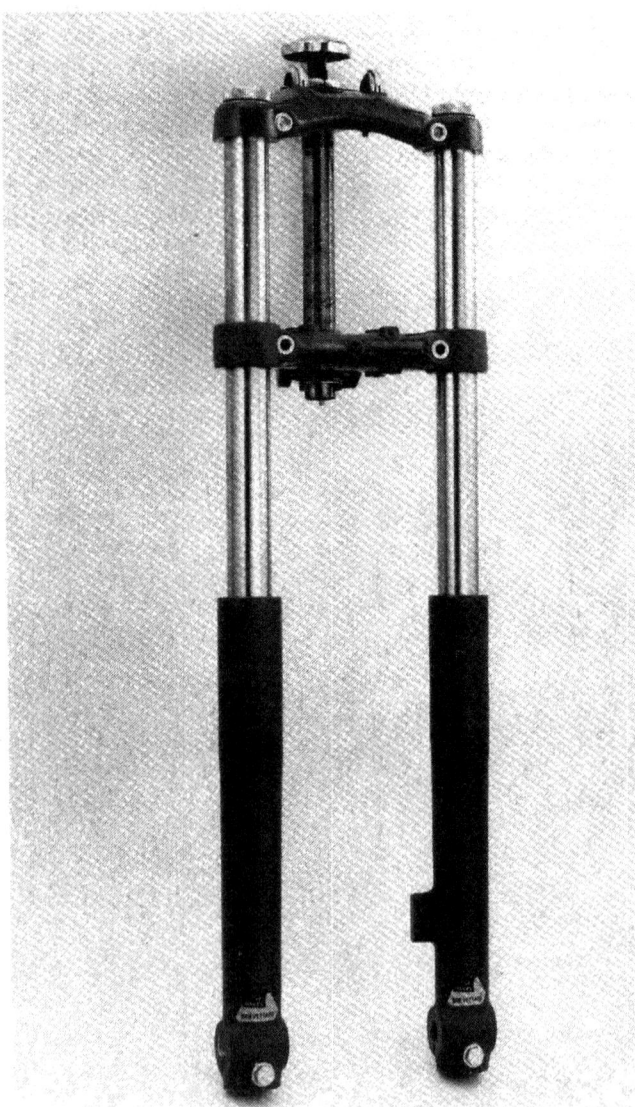

Early (for our purposes at least) Ceriani forks and clamps came over with the Aermacchi effort and were quickly (1962) adapted and approved for the KR and KRTT. Steering damper (knob at top of fork stem) was built in. Harley-Davidson

Changing faces

Meanwhile, Joe Leonard transferred from motorcycles to cars, Indy cars in particular. He got a good ride with J. C. Agajanian, sportsman and owner of Ascot Speedway among other enterprises, and Smokin' Joe went on to become the only man to date who's been national champion on both two and four wheels. (John Surtees, the English racer, is at this writing the only man to have done so in world championships.)

Len Andres asked Brad to retire, to run the family dealership so he (Len) could retire. Brad did. But when Len's retirement allowed him to tune for other riders, Brad persuaded Len to do another machine for him, Brad, to make a comeback with. It worked. Brad won Daytona in 1960, second man to win that race three times, and he won the national at Watkins Glen in 1960. But it was work. He said later that, like Resweber, what had once been instinctive now required thought. He now had to plan what to do, while before he'd simply done it. Not good, so Andres retired permanently from racing and went into real estate and flying. Len, we'll see again.

The rider in Resweber's sights on that day of his big disappointment went on to win that mile race, and to win the title plate in 1962. His name was Bart Markel and his time was slightly different.

XLR

During the debate over the limit on compression ratio, and when the pro and con factions argued

Another experiment. This photo, taken in 1966, shows the conventional Harley twin tomahawks and cradle for the engine, but notice the twin backbone tubes, very much like those of the Norton Featherbed frame. The front and top tubes crossed at the steering head, also a Featherbed feature, and the top tubes extended back beyond the seat mount and the junction with the rear tubes to just above the rear hub, where the shock mounts were, and where the extensions were triangulated with the lower mount. This frame wasn't produced. Harley-Davidson

Model	XLRTT
Year	1958-68
Engine	45° V-twin, iron barrels, ohv
Bore and stroke	3.00x3.8125 in.
Displacement	54 cu. in. (883 cc)
Brake horsepower	77 (1967)
Transmission	4 speeds
Wheelbase	54 in.
Weight	350 lb. (dry)
Wheels	19 in.
Tires	4.00
Brakes (factory)	8½ in. drum f/r

about what exactly is a racing bike and what's production, Harley-Davidson was careful to point out that the KR, KRTT and KHR were still in the catalog.

They had to remind people of that because the factory's actual sport machine was a lot unlike the KR or KRTT.

In brief, the stroker KH had been faster and thus better than the original Model K, but it hadn't done as well as Harley-Davidson had hoped. Just as Chevrolet went from six to V-8, and as Ford went from flathead to overhead, so did Harley-Davidson go completely modern, with the legendary, evergreen, everlasting Sportster.

The whole project was pure Harley-Davidson. The Sportster, using engine code X, then XL as compression ratio was bumped a notch, was the same frame, wheels, brakes, suspension and so on as the KH. The XL engine was the four-cavity unit design,

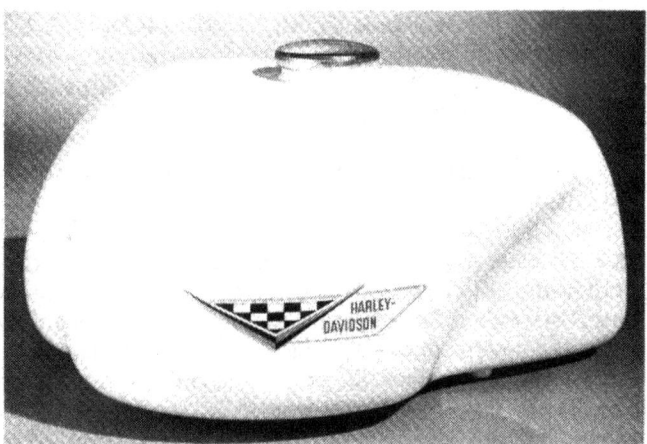
Optional 6 gal. tank was as big as the riders liked and seldom was seen away from Daytona's 200 mile race, although H-D offered and got approval for larger tanks. Harley-Davidson

V-twin, fork and blade rods, four speeds, foot shift, chain primary, 6-volt ignition.

The difference that showed was overhead valves. Because the KH was a 55 cu. in. rounded to 883 or 900

Highboy frame, 1968, with shocks and swing arm attached. This was a mixture of the 1963 and Featherbed frames, with a wide U for the top extension and the rear tubes tipped back slightly. The steering head was lower on the front tubes, which overlapped the backbone. The backbone was level and the head steady was closer to it, indicating the backbone had been brought lower, closer to the engine. Harley-Davidson

Oldani brake, another benefit of the Italian connection, had four leading shoes. This version, approved in 1962, was 250 mm in diameter. Harley-Davidson

cc, and the XL was a 55 cu. in., or 883 or 900 cc, and the engine could share gear case, primary covers and a host of other parts, the books generally assume that the XL was the KH but with overhead valves.

Not quite. The KH was a stroker, with the K's 2.75 in. bore but stroke increased to 4.56 in., from the K's 3.81 in.

The XL had the K's stroke of 3.81 in. but the bore was increased to 3.00 in. This gave the same displacement but lower piston speeds, so the engine could be revved higher, safer and longer. And because the valves were above the piston, the larger the bore the larger the valves. Therefore (other things being equal), the engine could flow more air and make more power, which it did then and has been doing ever since.

Back to tradition. TT rules and some hillclimbs allowed engines with more than 750 cc, so the KHR was replaced by the XLR.

Strictly speaking, and in keeping with the Harley system, the new production-based Class C bike was the XLRTT, as it came with brakes and rear suspension. There was an XLR, with rigid rear and no brakes, deserved only in the factory photo file. Clyde Denzer says to the best of his knowledge that rigid XLR was the only one built. None were offered to the public, he says, and no rigid XLRs were ever sold.

That makes sense, just as the specifications make sense, and for the same reason: The XLR was a TT bike. And it was still another perfect example of a Harley-Davidson production-based racer.

The XLR engine used cases like those of the XL Sportster, except that the main bearings were ball bearings instead of roller. The crankpin was the larger one, from the KR, but the rods were unique to the XLR engine, as were the flywheels, because the mainshafts were different. The XLR got the quarter-speed oil pump and the magneto in the front of the gear case,

Fairings were approved for road racing in 1963, so this is a 1964 KRTT with the factory-offered fairing kit. Otherwise, it's an old-style customer machine, with the sprung saddle, fender pad and Sportster-style brakes and forks. The rear top frame junction—the seat and shock mount— is a casting and the rear brake is integral with the sprocket. The 6 qt. oil tank was normal for road racing. The small panel on the right side of the fairing, just under "-Davidson," removes to give access to the front spark plug. Harley-Davidson

like the KR, while the camshafts were different from the KR but could be shared and swapped with those from the XLCH, the hot street Sportster. Clutch, kick start, primary drive and internal gear ratios were all shared with the KR.

So was the frame, except that the only frame listed was the KRTT frame, with the swing arm pivoting from the twin tomahawk and the shocks bolted to the rear of the seat casting. The rigid rear section wasn't listed for the XLR, nor was the light mainframe, the one built with thinner wall tubing for KRs.

There was a concealed handicap built into the XLR but it went unnoticed for several years.

The XLR was made of iron (as in the cylinders) and the heads were cast iron atop alloy cases. Iron is good stuff, useful and valuable, been around for ages. And alloy was nothing all that different for Harley-Davidson. Engine cases had been made of aluminum for generations. The big twins, the Panhead at least, had alloy heads on iron barrels on alloy cases. The KR heads were aluminum and the barrels were iron.

The XL engine, the Sportster, came after the Panhead but stuck with iron heads. Possibly it was cost. Iron is cheaper to buy and to machine and cast. Harley management had rejected an overhead cam V-twin with wider included angle when it approved the plans to base the XL on the KH, so perhaps it was simply a matter of money that put iron into the XLR.

At any rate, the XLR engine used standard XL barrels and heads that were reworked, legally of course. The XLR got larger ports and valves and the spark plug boss was machined down for a different plug. The XLR used Linkert and, later, Tillotson carbs, same as on the KR engine from the same time.

As with the KR, the XLR was a matter of evolution. And there were various claims made. At the time, O'Brien was willing to tell the press everything he wanted his opposition to hear. Thus, news accounts of the period claimed 80 bhp and up, right from the start, while now from his retirement haven O'Brien says that the standard XLR—with 1 15/16 in. intake valves, 1¾ in. exhausts, the mid-peak PB cams shared with the XL, and with stock ports and carb—had 77 bhp, in 1967 customer form. The earlier customer versions had less and the later engines, especially when tuned by skilled people, had more.

The XLR—I'll skip the TT part here, because there was no other version—never got the recognition it deserved.

The rules worked out conveniently, in that the XLR, in the form of the Sportster, was recognized for Class C by the AMA without a lot of fuss. Racers were already using the KHRTT for TT, since the old 74s and 80s were hopelessly outgunned by the Triumph and BSA 650 twins.

The TT rules changed in 1963, with classes for lightweight and heavyweight, with limits respectively of 250 and 900 cc. This was partially inspired by the new classes in short track, with less restrictive rules but with a 250 cc engine limit, and partially by the logical disappearance of the really big machines. Perhaps in some small part, by making sure that the Harley engine, 900 cc when bored out, was going to be the largest legal engine in the class. The Vincent, still in the record books although out of production for ten years, was 1000 cc.

The XLR was a monster. It had stupendous power, more than the KR or KHR could ever hope to have, and it was only a bit heavier than the flathead machines. The brakes were no better than you'd expect from part-time equipment and at the time the tires were designed *not* to grip the track too much.

The XLR took a special kind of rider. The two best, from the record, were Joe Leonard, who was big and strong and held the record at Ascot long after he and the XLR had been retired and replaced by more modern equipment, and Mark Brelsford, who was about six times stronger than he looked.

"He made me work for my money"

The speaker was Carroll Resweber and the subject was Bart Markel, sometimes known as Black Bart. Markel hailed from Michigan and ushered in a different era, along with his peers.

Dick O'Brien's job wasn't easy. As head of the Harley-Davidson racing program and shop, he was supposed to provide parts, support and technical advice for racers who were outside contractors. Markel and, later, Mert Lawwill were the most successful but had to contend with Harley guys Roger Reiman,

Mert Lawwill's KRTT, 1965, set up for a short road race, minus fairing. This old-style frame had forward mount shocks and conical (Sportster) front brake. But the rear brake, larger, was on the left. That's the 3.7 gal. fuel tank, with a plastic seat base and tail section. The right peg was moved so far back that the shift lever was trailing instead of leading (and the shift pattern was reversed, which could get confusing in the heat of battle). Harley-Davidson

whose family had a dealership; George Roeder; Len Andres' new rider Ralph White; and a field of privateers.

And racing has always had the loyal opposition. Somehow, for every bright, tough Harley racer there's been somebody just as bright and just as tough riding for Indian or Excelsior or, in this case, for BSA; one Dick "Bugsy" Mann.

Markel was tough; he'd been a Golden Gloves boxer. Legend has it that another racer was supposedly not just a bad man, he had a bodyguard and the bodyguard supposedly had a black belt in one of the martial arts. Well, push came to shove and Markel decked the dreaded martial artist.

Another time, some rowdy outlaws in the stands gave Markel a bad time, so he jumped the fence, steel shoe in hand, and laid waste to the gang.

Nevertheless, Markel was an exemplary family man and a careful planner. And he was fast.

Just as when Resweber ran out of gas and Markel went on to win the race, so did Markel take over the points lead after Resweber's crash and go on to win the national title for 1962. He also won six nationals in 1962, so it wasn't merely being in the right place at the right time.

The outside contractors got to build their own equipment. Bill Milburn, a former racer who restored bikes from this era, bought Lawwill's TT machine years after Lawwill retired and was puzzled by a linkage that seemed to run from the rear brake pedal to the clutch worm, the skew gear that pushes the clutch release rod in.

Then, the light dawned. The XLR used lots of compression and not much flywheel. Sometimes in the heat of battle a rider could stomp on the rear brake and lock the wheel, which killed the engine. So Lawwill had devised a linkage by which, when he stepped on the rear brake lever, he disengaged the clutch, hence no risk of killing the engine in mid turn.

O'Brien remembers the time Markel lined up for the main event with a tire O'Brien knew wouldn't work—wasn't the right compound to go the distance. Markel knew it, too. But he figured it would give extra grip the first laps, enough to let him pull ahead, so even if he lost grip later he'd be well into the lead.

Off they went, and sure enough, Markel grabbed an early lead. Then his tire went sour, he blocked the other guys when a narrow groove developed on the track, and he won.

In 1963—sort of a forecast in that there were fourteen nationals and Harley-Davidson won seven—there were so many good riders that no one man dominated and Dick Mann edged George Roeder by one point while Markel scored a single win, a light-

By 1966 Roger Reiman had gone to the short modified frame and long and low tank and seat. The front brake has been drilled for cooling, the shift has been reversed and the factory fairing lowered. Daytona Speedway

Remember the factory's version of the Featherbed frame? That wasn't the only one. This is Ken Ridder, winner of the Junior race at Daytona in 1964, with a KRTT out of Dud Perkins' shop. The frame was by Jim Belland and had two top tubes extended back to the shock mount, with curved tubes from the seat junction down to the rear engine mount. The different frame took a different oil tank, below the seat. Daytona Speedway

weight TT. Roger Reiman edged Mann in 1964 and Markel came back, comfortably ahead of Mann in 1965 and Gary Nixon, another sharp racer who rode against Harley-Davidson, in 1966.

Competition off the track

But that's merely the scoreboard. Behind the scenes, so to speak, there were several sorts of pressure being exerted on (against?) Harley-Davidson. In an innocent way, road racing was getting more grassroots support, with non-AMA clubs running events on road courses with international style rules—many displacement classes, no design limits and full fairings. Fairings were banned by the AMA, mostly because street bikes didn't have fairings but also because the Class C spirit meant that racers could run all the events with one bike. They couldn't do that effectively from the beginning; try taking the brake internals off, putting them back on, then taking them off again week after week.

So there were flat-track bikes and road-race/TT bikes. Short track was revived as Class A in the mid-fifties; then in 1959 the short-track national races were limited to 250 cc, with virtually no other limits. At first these races paid half the points you got in other nationals, but that was changed in 1961. So there were three machines, the 250 short-tracker, the 750 (or 500) flat-tracker and the road-race/TT bike, needed by all those who wanted a shot at the championship.

With this specialization came the ability to further focus the purpose of the machine; you could build a better miler if it would never have to run the short tracks.

This made for some confusion in some, uh, circles. BSA advertised that its guys were the Class A champs, which they were, sort of, in that they won more races, even if there wasn't a Class A championship to win. And another BSA ad said "Fastest motorcycle at Laconia! Dick Mann and his BSA Gold Star led the 100-miler for over 90 record breaking laps!" Which they certainly did do. What the ad didn't say was that the race ran longer than ninety laps but the BSA didn't. The Europeans may have laughed to see Brad Andres cornering with his foot down, like the dirt guys still do, but the KRTT ran the 100 miles and the BSA didn't.

The coin flips again, though, as the 500s and the imports get better and their options increase, and the pressure for changes gets stronger. In 1959 the AMA Technical Committee voted 25 to 3 not to allow brakes in flat track. In 1961 the committee approved a test, the use of brakes at Ascot, J. C. Agajanian's short, half-mile and TT track combination south of Los Angeles.

There were other, even odder, pressures. The AMA races at Willow Springs may have revived road racing in America but the non-AMA races continued, a direct parallel to the sports car versus oval track feud on four wheels. The AMA didn't belong to the International Motorcycling Federation (FIM), so up sprang a club that did join the FIM, and put on its own US Grand Prix. The guys in that club said no American could hope to compete with real road racers, causing Floyd Clymer to fire back with a challenge: Leonard and Andres to go to the Isle of Man with two "K competition models" (Clymer's term) to run against any two British riders, with the clause that neither Brit was to have ever seen the Isle of Man course before, so they'd be on an equal footing with the Yanks.

Right. From a historical viewpoint, this all sounds remarkably foolish. Not only that, Clymer, in his haste to defend honor, moved so fast he forgot to ask Leonard and Andres what they thought of the idea.

Not much, as you'd expect. Leonard simply wasn't interested, period; he was thinking about other things. Andres, the fierce competitor hidden behind the shy demeanor, said much later that first, the Isle of Man is the very last place in the world for such a match race and, second, he would have liked to accept the challenge if it had been fair. In fact, a few years later he got an offer to race in Europe but he wasn't sure he'd have a first-class machine, so he declined. Still later Mike Hailwood, who won the non-AMA US Grand Prix at Daytona, was invited to race against the AMA guys. Amiable to a fault, Hailwood said he'd like that except that he belonged to one club and the AMA was another. He rode for a factory that had rules; in short, no thanks very kindly.

Thing here is, there were strong feelings, pro and con, on both sides. Love-Hate, that's the term. But deep down, everybody probably knew that you can't have one motorcycle that's good at both dirt-track and road racing.

And, in 1960, one year after Mann led those ninety record-breaking laps at Laconia, Mann won the race, ending a seven-year Harley streak. In 1961 the AMA's Daytona race moved to the speedway, but carefully laid out a flat road course in the infield. No bankings. The United States Motorcycle Club (USMC), the world-affiliated club, ran on the bankings. In that 1961 USMC race, a streamlined Matchless won and the best Harley was seventeenth. In the AMA's Daytona race, Roger Reiman was first on a bare KRTT, with a Triumph second.

Here we are at 1962, the turning point. In the political arena, the racing Nortons and Matchlesses were approved for Class C. Quoting *Cycle*, "We are undoubtedly going to see better racing now that the AMA has let down the bars."

At Daytona, Resweber, Leonard, Reiman and Roeder were the fastest qualifiers. And a Triumph Trophy, ridden by Don Burnett from Danvers, Massachusetts, outlasted the Harleys and won. It was Burnett's only national win, but the point was made.

On the technical front, more or less in sequence, Harley-Davidson submitted and the AMA approved, a

George Roeder (94) was also on the Harley factory team and had a KRTT with normal bodywork done up in wild curliques. The front brake was the big Oldani, with scoop, and the fork brace was a straight bridge across the legs. Team riders had lots of freedom in working out the details. Harley-Davidson

7.8 gal. fuel tank for the KRTT, and four-leading-shoe front brakes for the KRTT (and the CRTT, the lightweight described later). Triumphs got fiberglass tanks but the Bultaco—no model listed—wasn't approved because the papers weren't in order.

The claiming rule, by which one rider can file and get the other man's machine, was set at $1,200. This was another way costs were supposed to be kept

Harley's semi-team approach, illustrated at Daytona Beach, 1966: That's Mert Lawwill (18) with the standard fairing and fiberglass tank and seat, but the forks had an arched brace, like those used in desert racing then, and the front brake had cooling holes but no scoop. The paint was sedate and the emblem on Lawwill's leathers advertised Dud Perkins, out of whose shop Lawwill raced.

down: If you could be forced to sell your bike for a set price, why, you wouldn't spend any more than that building the thing. In fact, most guys did spend all they had, claiming rule or not, and because they all loved the results of their work, they didn't claim each other's machine unless it was for personal reasons.

The national TT class was changed from 45 cu. in. or 80 cu. in. to lightweight or 250 cc, and heavyweight or 900 cc.

The committee opened the meeting to comments from the floor, and there was a discussion of brakes for dirt track and streamlining for road races.

Streamlining, that is, fairings, were approved for experts in 1963. Brakes were turned down, again.

Try a new frame

Next is the sort of event that should be announced in large letters or bold-face type: The factory goes racing. On the other hand, the events here are fairly subtle. The baseline fact was that late in 1962 the AMA's Technical Committee approved a new frame for the KR, KRTT and XLRTT. This frame needed committee approval because the production parts rule was still in effect.

We need to break this major event into two parts. The first part is going to be with us until we run out of petroleum. With this new frame, road and dirt-track machines began to differ. Okay, you probably really hadn't been able to use one machine for both types of racing since the demise of springer forks, but this was the beginning of the total separation.

The second part is a sort of legal and moral issue: The frame submitted by the factory and approved for use in racing on the explicit grounds that it would be available for sale to any racer never got into the parts books.

Harley-Davidson, while offering all help possible on engine tuning and preparation, was keeping this one for its own use, for the racing shop to provide to the best guys. The parts books issued later have the same old numbers, for the frame with the cast-iron steering head, rear top section, and twin tomahawk for the rear subframe or the swing arm. This conclusion is confirmed by Clyde Denzer, who said years later that the factory would sell you a kit, such as the KR frame, the swing arm and the fairing, but that the customer always had to build his own road racer.

This is the back end of Bart Markel's Lowboy KRTT, at Daytona 1968. The rear exhaust pipe was kinked out to clear the shock/spring, and the triangular oil tank tucked nicely between the rear frame tubes and the seat base. The rear of the frame backbone and the brace from backbone to the steering head were wrapped with tape over foam, for padding. Harley-Davidson is supposed to have been the first to use that pointed, streamlined tail section. Daytona Speedway

For Daytona 1967, Cal Rayborn had an Andres-tuned KRTT, with fiberglass tank and seat, plus a new fairing. But the frame was early Highboy (witness the seat extensions back to the shocks), while there was room below the seat for the old half-circle oil tank. The foam pad just below Cal's right forearm was glued to the tank so he could rest his chin on the pad down the straights. Front brake was the four-shoe Ceriani. Rayborn had top time with this bike but the engine lost a main bearing and a Triumph won. Daytona Speedway

This moral shortcoming didn't stay in effect for long, though.

A technical digression: Motorcycle frames have never been the science the builders and designers would have you think. Frame design instead goes against logic. Racing cars have for generations been built with the idea, which is logical, that the stiffer the frame, the more predictably the suspension will work because it will have a stable platform to brace itself against. And for fifty years or so, cars have gotten stiffer and stiffer in all their dimensions.

Not motorcycles. Back in the fifties two Irish brothers built some racing bikes with twin loops of tubing all the way around the engine, and with the front tubes running to the top of the steering head, crossing the backbone tubes which went to the bottom of the steering head. The lucky riders, so equipped, said these bikes "rode like a featherbed." And so the design, which appeared first on Nortons, was known as the Featherbed frame. The design worked and was copied worldwide, although nobody could really prove why or how the idea worked.

Twenty years later, in the seventies, English theoreticians and Japanese engineers, working separately, made public some of what they'd decided. Both camps said, summarized and in broad outline, that the major factor in making a motorcycle handle is keeping the front axle under perfect control, and that nearly as important is having a frame with optimum stiffness in torsion.

Note those qualifiers. The first part, the axle, seems reasonable. A controlled axle keeps the front tire's contact patch in control and the rest of the bike follows along.

The frame, though, takes some thought. There is no limit to how stiff the frame should be in compression or tension, between the two wheels or where the wheels connect to the frame, or between pivot point and steering head. The stiffness limit there is imposed only by weight and space.

But torsional stiffness is optimum. Torsion here means twist, the loadings fed into the frame from cornering, braking, steering and engine torque, all at once and in several directions. There seems to be a need for some flex, damping in a way, in these dimensions; but not much. Easier to say than build into a racing motorcycle.

Little of this was articulated, that is, the guys didn't sit around the pits drinking Cokes and worrying about optimum torsional stiffness. Instead, the smartest and most ambitious began working on frame design for themselves. And, according to a survey taken in 1963 by announcer Roxy Rockwood, the

An early Lowboy frame, as in late 1967. This one had the standard twin tomahawks for the engine mount and swing arm pivot. It was all tubing, but the rear section was extended into a triangular brace/mount/support for the rear shocks, with smaller extensions for the cone rear seat section. There was the main backbone tube, virtually horizontal, with a brace from its center to the steering head. The boss for the head steady shows how close the backbone was to the heads. (The four bosses at right front are the tubes through which the engine mounting bolts passed.) The strap across the rear U section held the seat and fender, with the oil tank below that and tucked in the vee formed by the center and rear downtubes. Harley-Davidson

Winner Rayborn, who looks as if he already knows the Daytona 200 is in his pocket for 1968. Those were the road-race Ceriani forks and four-shoe brake, the then-new Wixom fairing, and the full-team color scheme of orange and black with white. Daytona Speedway

rider wanted a free choice of frames, along with more money and an end to the ownership rule.

Also part of this interest in frame design was that while stress had increased all around, there was more power from the engine and more traction from the tires and track surfaces, so the link between the two—the frame and suspension—had more to do. And road and dirt racing were becoming more and more different.

Skipping over most of the theory and nearly all of the illustrations, in general, the road and dirt machines needed different weight distribution, center of gravity, wheelbase and steering head angle (rake) along with the brakes/no brakes division.

And of course when this major revision began, in the early sixties, there was still the choice of rigid versus suspended rear wheel.

This was a choice, because all the rules said was that you had to show up with an approved, accepted motorcycle. The AMA didn't care if it was a KR for road racing and a KRTT for the mile. But when this began, the rigid frame was much better on the smooth, cushion tracks of the day. The advantage was so obvious that BSA and Triumph offered rigid frames, for America only and for racing only.

Then the balance began to shift. O'Brien says it was mostly the work of Dick Mann and Mert Lawwill, who thought about racing twenty-four hours a day. Mann rode BSA or Matchless, so he had the factories' samples on hand, as it were, while Lawwill says he built up his first springer frame in 1964, before the

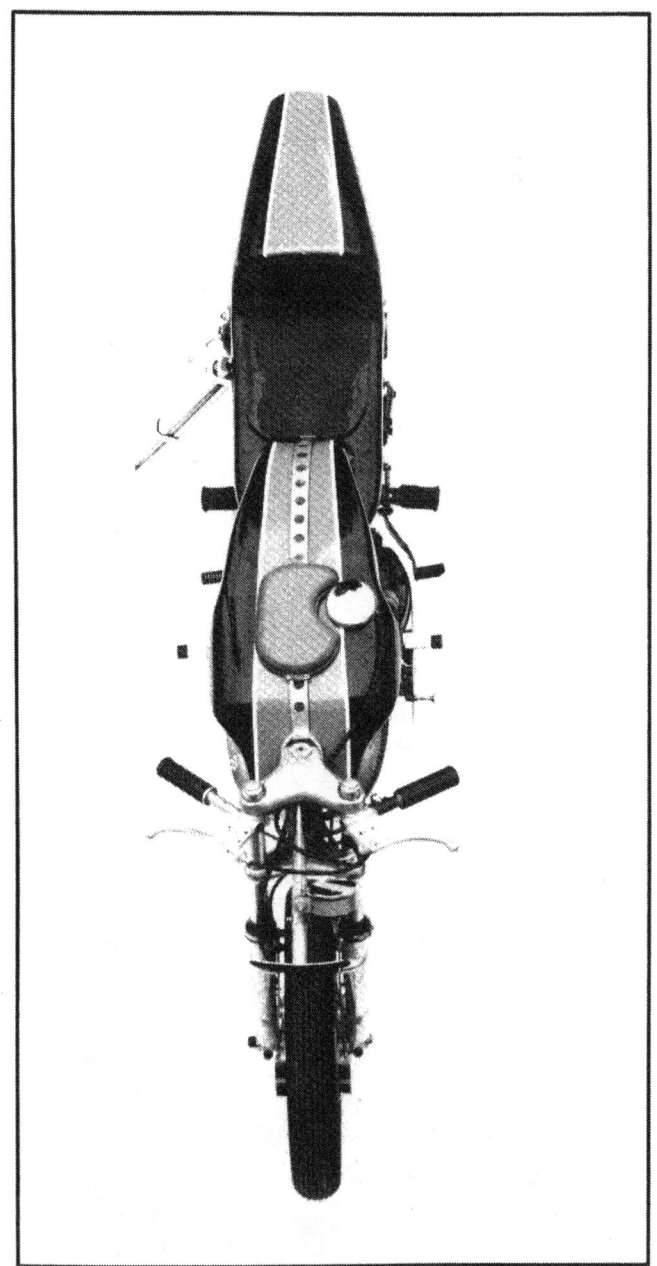

Narrow is what works for V-twins. This is a 1968 team bike, shown from above to illustrate just how compact this machine was. Notice the clip-on bars, the exhaust pipes tucked below the seat and the pad for the rider's chin on the tank. *Harley-Davidson*

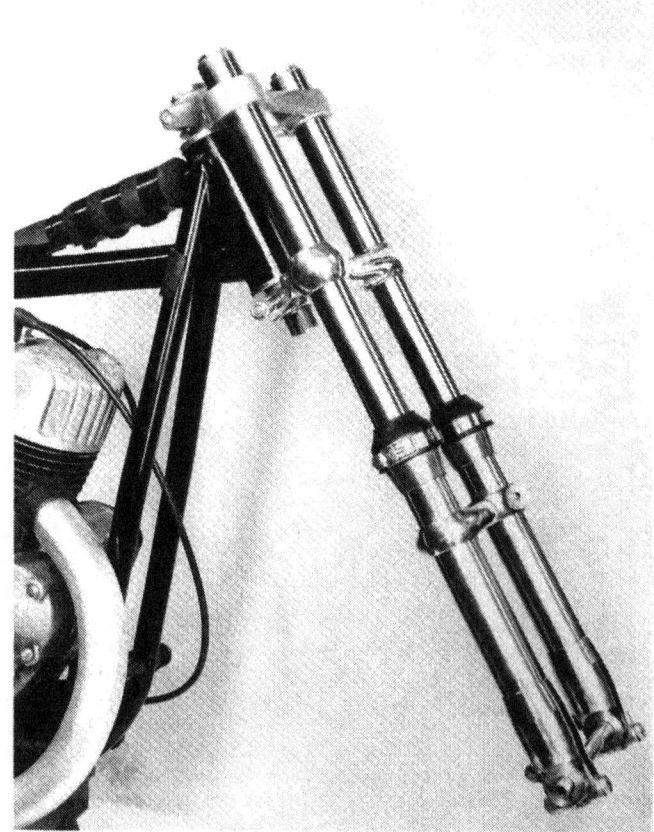

Same vintage frame, with horizontal backbone, fitted with the Ceriani clamps and forks. Extra length was built into the stanchion tubes (the pipes gripped by the clamps) so the front end of the bike could be raised and lowered by sliding the tubes up and down. *Harley-Davidson*

Harley factory was talking about it. But the change took a while, because the frame builders had to work out spring rates and find shock absorbers that worked.

Test results

Meanwhile, the technical inspectors rejected Dick Mann's Daytona entry for 1963 because the frame hadn't been submitted to and approved by the Technical Committee. He said, with some justice, that he'd raced the frame all through the 1962 season, so he'd figured it was legal. The AMA said, also with justice, that it had told him before the race that he couldn't run that frame; he could have built a bike with a legal frame, but he didn't, so he was out.

Harley-Davidson won the battle, as Ralph White, riding for Len Andres on a KRTT set up by Clyde Denzer, won Daytona. Mann won the war, edging Roeder by one point for the national title. White was third. (At the time, the points system gave one point for every rider you beat, down to 30, so road racing with its larger fields paid extra compared to the dirt tracks.) What really counted was consistency, as Mann won only one race that year, the Ascot TT, and got his points by nearly always finishing near the top. Roeder won three races but didn't do as well week after week. And Harley won half the 1963 nationals, with the imports taking the other half.

At the time, the top riders, those with factory backing, were their own best tuners. Markel and

O'Brien's finest hour, Daytona 1968: It was the first lap of the big race. Off the straight and into the infield came Mert Lawwill, followed by Fred Nix, Cal Rayborn and Bart Markel, with Roger Reiman on the outside and Dan Haaby close behind. Harley-Davidson

Lawwill ran their own programs, with factory help in money and advice.

Some accepted help more readily than others. Bill Milburn dug through the factory's engine room records and found that it had shipped Markel twenty

The Harley-Davidson Team, as a team: equipped with KRTT Lowboys, the then-new fairings and the team paint scheme before Daytona, 1968. From left, Bart Markel, Mert Lawwill, Dan Haaby, Cal Rayborn, Roger Reiman, Fred Nix and Walt Fulton, Jr. Harley-Davidson

A road racing Lowboy frame, covered with tape for padding. The rear (seat) section was low, then the backbone arced up to clear the head and down again, with a brace to the top of the steering head. Ted Pratt

sets of engine cases, all with Markel's ownership number on them.

Markel was also stubborn, O'Brien says, "You'd have to kill him to change his mind." Sometimes (witness the tire incident), Markel was right, sometimes he was wrong. When that happened, O'Brien says, Markel would admit it.

Lawwill was much the same, in O'Brien's view, and he cites the time Lawwill had been winning with one combination and showed up for a national with a different frame, one that turned out to be hopeless. The good bike was home in the garage. O'Brien wondered why it hadn't been hauled along, just in case, but Lawwill said he'd figured the new one would be so much better he hadn't bothered.

The payoff at the time came slowly. Harley-Davidson had to develop new frames and improve an old engine, one that by then everybody knew had to be on its last legs.

The KR's death, like that in the Mark Twain anecdote was somewhat overreported. O'Brien had a good shop and proper equipment at the factory. Tuners C. R. Axtell, Jerry Branch and Jim Belland were working outside, independent but sharing.

Exactly what they did and how well it worked is tough to trace years later because of the psychology involved. *Cycle World* did a track test of the 1963 KRTT in 1963. With fairing—ninety percent of the bikes at Daytona that year were streamlined—and with the six gal. fuel tank and 3 qt. oil tank filled, the bike weighed 386 lb., one heck of a lot for a racer and more than earlier reports predicted.

That engine was rated at 48 bhp, claimed, at 6800 rpm and the magazine said the KRTT would hit 142 mph, revving to 7000 rpm with the gearing installed for the test. In 1965, *Cycle World* said that the KR engine had 52 bhp. In the same time period *Cycle* said the KR would crank out 58 to 60 bhp toward the end. The press reported 50 bhp at least from the lighter British engines, so power-to-weight, as shown years

And at the finish it was Rayborn, the lone survivor of the team, making up for that by lapping the field and turning Daytona's first 100 mph average. Harley-Davidson

earlier in Harley-Davidson's test for the AMA, was probably just about equal.

That both camps were talking stronger than their engines actually produced, is a fair guess here.

They had the same rewarding bottom line, though, in that there were close, fair races and multiple winners. Roger Reiman, the Illinois dealer and a man the AMA always suspected of shifting during races but never managed to catch, won the championship in 1964. Reiman got his boost from Daytona, where he won. His other win that year was a short track on a lightweight Harley Sprint. Mann won four races and came in second on points, Markel won three and came in fourth, so consistency paid off again, but it paid off for all parties. That year there were seventeen nationals and Harleys won nine of them, just over half.

In 1965 Reiman won Daytona again and Markel won the No. 1 plate again, as well as three nationals. Dick Mann was second in points but got three more race wins.

And Harley-Davidson won only six of the eighteen nationals.

Technical Committee votes

Longshoreman and philosopher Eric Hoffer wrote that revolution doesn't cause change, change causes revolution.

Such proved to be the case in Class C. Not that anybody noticed (or predicted) when the revisions began. Instead, the 1963 Technical Committee meeting saw a vote of 23 to 3 to keep the equivalency rule, a 20 to 6 vote to change short track from Class A to

A bit out of sequence, this photo was taken at Daytona Beach, 1985, and shows what the vintage guys will do. That's an XR frame and disc rear brake, an alloy XRTT fuel tank, an accessory half (or maybe even quarter) fairing, clip-on bars like the factory never used and a Suzuki front wheel, all propelled by an authentic KR engine.

Class C (there now being enough 250s from Triumph, Honda, Harley et al to make it practical) and to allow use of the Daytona banking, 22 to 4.

Brakes were still banned on dirt except for novices at Ascot, 23 to 3. Streamlining for all road race classes was approved, 23 to 3. The G50 Matchless frame was not allowed, 15 to 11. The tattered old ownership rule was finally abandoned, 19 to 6. And Class C frame rules for road racing were retained, 16 to 8.

Racing was booming; 1,800 riders had Class C licenses, three grades, and 800 Experts expected to contribute to the new points fund. (Ninety-eight riders earned national points in 1963.)

The committee was reversed when the dreaded Matchless was later accepted. There was a motion in 1964 to limit Class C to 350 cc. Rockwood said the riders hadn't been asked. One assumes there was some sort of deal, because the public wasn't buying the 500s the English raced and Harley wasn't even building, for public sale anyway, the 750s that Harley raced. Neither side had an engine the other side could compete against, so the one they did share, the 350 cc four-stroke, could have been a compromise.

Norton got approval for large Oldani brakes and Harley-Davidson did the same for alloy brakes. In another tit for tat, Triumph's racing manager, Rod Coates, and tuner, Cliff Guild, defended their factory's racer: The Triumph T100S/R was a short-stroke, big-bore version of the Tiger T100, with oversize intake valves, special exhaust valves, radical camshafts and ported heads. The latter were available to dealers on an exchange basis and booklets with all the parts numbers were offered to all comers, for the asking.

Roger Reiman, still in sequence since he rode his KRTT in the 1985 vintage race. It was at least as fast as ever.

The shoe was pinching both feet, to rework a phrase.

But that was a sidelight. Former executive director E. C. Smith had retired. New director Lin Kuchler appointed a new Competition Committee. William Harley moved that licensed racers be represented (a radical idea), so among the members were Reiman and Mann.

By a vote of 23 to 3 the committee decided to abandon the old 500/750 formula. Instead, Class C limits would be 750 cc, no more than one overhead camshaft, two cylinders, two valves per cylinder and four speeds forward. This would apply to all Class C professional races except short track, and it would be subject to reconsideration in 1965, while not taking effect in any case until 1967.

In other actions, gear shifting was permitted, 22 to 4; tire treads could be trimmed and cut, 26 to 1, brakes were still banned from dirt track, 14 to 13; and the requirement for Class C eligibility was changed from twenty-five to 100 machines, a switch that will come back to haunt the sport.

Reiman won the 1965 Daytona 200, over the full infield and banking course, with Lawwill second. Of the first twenty-five finishers, nine were Harleys, eight Triumphs, four BSAs, three Matchlesses and one Norton. Markel won the national title again because he was strong on dirt and consistent on pavement.

Another odd machine, even more proof that there's no such thing as a stock racing motorcycle. Also from the vintage circuit, this KRTT had the dual carbs and disc rear brake from 1968, but the Ceriani front brake. And the frame was made by a St. Louis dealer.

Harley-Davidson got six national wins, Triumph five, Matchless four, BSA two and Yamaha one—a road race in which the 350 cc two-stroke beat the 500 and 750 cc four-strokes (that was another harbinger few noticed at the time).

The minutes of the Competition Committee meeting in 1965 take some careful study. The previous year, remember, members had voted for a fairly straight-across 750 cc limit, with no multiple valves or cylinders. They'd also agreed to think about the matter again.

So this year they thought again... and rescinded the 750 rule. The vote in 1964 had been 23 to 3 for; now they voted 21 to 4, against. Next, the Classification Committee submitted a proposal that the present (that is, 500 ohv 750 side-valve) class be kept until 1969, when it would be replaced by a class requiring a 350 cc limit, production engines only, air cooling only, no more than five speeds, no supercharging and 200 examples in the United States before approval.

This one passed, 22 to 2. Then there was a motion to keep this rule in effect until at least 1971, because the quick changes were confusing everybody. (They certainly were!) This passed, 22 to 3.

The word here is *politics*. Reliable sources—people who were there but don't want to go public—said that everybody wanted to keep down costs and speed, and nearly all the factories could come up with a 350, while 750 was too big for the English, too small for Harley.

Then it was back to normal business, discontinuing the Ascot brake experiments, and once again prohibiting brakes in dirt track, 15 to 8.

There was a motion that road-race frames "must be the same as submitted and endorsed by the manufacturer and/or the U.S. distributor of the motorcycle to which the frame is distributed... Photos of installations and specifications must be submitted." The motion passed, 25 to zero.

Highboy

That sounds fairly clear and concise. But the Harley-Davidson racing department knew it needed to do something. Buddy Elmore won Daytona on a very fast Triumph built by the Dallas dealership, so in the middle of the 1966 season some of the KRTTs appeared with a new frame.

The name became the Highboy, although it wasn't used at the time. The new frame was all tubing except for the twin tomahawk. The steering head was lower on the front downtubes, which crossed the single backbone tube in Featherbed style, and instead of a short junction for seat and rear shocks, there was a horizontal U-shaped extension, with shock mounts directly above the rear wheel and with triangulation from just above the swing arm pivot to just in front of the top of the shock.

That was the team bike, the one built at the factory and loaned to the best riders. No parts numbers appear in the books, however, and customers continued to get the regular frame with the short, cast rear section.

In 1963, quoted power for the KR engine was 48 bhp. For 1966 it had been inched up to 51 or 52, depending on the source. The factory was watching every ounce, such as trimming the lower panels of the fairings on short road courses where the lower weight would mean more than the extra drag. And Harley-Davidson had been joined by Fred Nix, an Oklahoman who statistically was the best mile racer of this period.

Even so, Markel kept the plate, again because he never gave up and was always near the front. But he won only two races in 1966, a half mile and a short track. Harleys won five of the fourteen nationals.

Lowboy

Motorcycles as a hobby, if not as a form of transportation, were growing in popularity. So was racing, as 2,500 riders held competition licenses, and some tracks, for instance Ascot, had to have alternate programs, with even-numbered Juniors racing one week, odd numbers the next. Because there were so many entrants it would have taken all evening just to run time trials without some limit on entries.

Harley-Davidson was sharing in some of the prosperity, but the company wasn't growing and the family owners were looking for an angel, or a partner, or even a public stock offering to raise money.

Even so, the Brits were so clearly hammering on the gates that O'Brien was able to invest in a development program *and* field, for the first time in generations, a factory team.

Instead of having favored riders use machines built by the racing shop and fictionally owned by the riders, O'Brien had a full shop, with machinists, fabricators and mechanics. He had five riders under con-

Neat rear brake was by Airhart. Just visible here are the lugs at the back of the twin tomahawks, the castings that identify a factory frame. Tubular swing arm looks like the ones still used by the factory. Daytona Speedway

tract: Markel, Lawwill, Roeder, a newcomer named Chris Draayer and a marvel named Calvin Rayborn (more about him shortly).

The 1967 season didn't begin well. Nixon won Daytona with Elmore close behind, on the Triumph T100R model quickly known as the Daytona. The Triumph factory team ran off from the Harley factory team. Roeder, Lawwill and Rayborn were third, fourth and fifth, respectively. They had been using short-rod KRs, remember them? And they had oiling problems. They also used cleaner fairings and four-shoe Ceriani front brakes, but none of it was enough.

Then, a loophole. O'Brien candidly admitted that when they first went racing with the Highboy frame, they'd wondered if they'd get away with it. They did, because first, the rule interpretation was based on whether the factory had approved the specs, which obviously H-D had done and, second, the frames were not *kept* from the private racer, they simply weren't thrust in his face. If you knew to ask for the new frame, the shop would build you one. (Nothing in the record says how long it would take to get one.)

Return for a moment to frame design and practice. By this time, dirt-track and road racing were essentially separate, with TT having moved to the dirt side. As suspension became more useful, the jumps got higher and longer and the TT bike needed more ground clearance. In contrast, the road racer was running on smooth surfaces, no oiled dirt and gravel, needed less suspension travel and thus could begin lower to the ground. With less wheel travel, the top could also be lower.

As a bonus for Harley-Davidson here, the US firm had acquired an Italian company, Aermacchi, makers of nifty little motorcycles. Details are discussed later, but the point is that H-D racing suddenly had access to good parts, like Ceriani forks and brakes. The competition parts were imported for the Sprint-based racer, CRTT, but were filed with the AMA as options for the CR and the KR; no extra trouble.

O'Brien and crew were hard at work, using whatever they could. The engine was nearly done, as it was taking more and more effort to get less and less improvement. But the frames were just at their beginning, in road racing at least.

The Highboy, the all-tube frame with rear section extended, ran Daytona with a newer and neater fairing. For Laconia, the suspension was lowered and Ceriani forks replaced the KR models.

The really new frame, known ever since as the Lowboy, made its debut at the Indianapolis road race national.

Low was only part of the story. The new frame wrapped around the engine, literally. There were two downtubes, from the steering head in front of and then beneath the engine, with the front mount a pair of alloy plates, just like the first K and KR. The new frame retained the twin tomahawks, the rear mount and swing arm pivot. And there were two rear downtubes, from tomahawks up to the wide U, the junction of the backbone tube and the seat/rear shock mount.

But the backbone was placed so it barely cleared the cylinder heads. The steering head was 2 in. lower than the earlier all-tube frame's had been, and the seat/shock mount tubes were 2 in. lower as well. With shorter shocks and fork tubes, the Lowboy KRTT had a seat height of 27 in., where the Highboy's was 30 in. Wheelbase was 54 in., the Highboy's was 57. (This measurement is arbitrary, as the wheelbase changed with chain adjustment in the rear, and with static ride height in the front, but the comparison holds.)

The battle continued

The 1967 season saw considerable experimentation. There were different shocks and forks, and the frames had several changes in wall thickness. At Indy, Roeder's bike had a much larger, 2 in. diameter, backbone. The Harleys ran full fairings at Daytona, bare at Laconia, half fairings at Indy and half fairings at Carlsbad.

At the end of the season, *Cycle*'s Gordon Jennings' KR, with backbone frame like that used on the built by tuner Jerry Branch, said it had been a fair year; Harley and Triumph were equal in power and handling, but Triumph had the edge in braking. Rayborn won one of the year's road races, Nixon won the other three. In a way that was racing luck because Nixon's Triumph qualified for Daytona at 134.238 mph, while Fred Nix, on the fastest KRTT, was clocked at 140.823 mph. The KR had the speed, it just couldn't sustain it.

Jennings defended the ill-fated short-rod motor by saying that it gave a wider power band, a gain of 2 bhp at peak and less loss of power from heat. Jennings' KR, with backbone frame like that used on the CRTT, weighed 310 lb. The KRTT with Lowboy frame

The KRTT's last famous victory—Cal Rayborn at Daytona Beach, 1969. The only difference from 1968 visible here is the megaphone exhaust system. Daytona Speedway

and brakes weighed about 325 lb. (One is tempted now to wonder if the KRTT ever weighed as much as the 385 lb. quoted in the various tests of the time. That figures seems 30 or 40 lb. too high, as if perhaps the team wanted the opposition to think the Harleys were heavier than they were.) The KR in 1967 was supposed to use a nominal compression ratio of 6:1, and have 55 bhp at 6300 rpm.

Things really were close. The season came down to the last race, Nixon and Triumph versus Roeder and H-D, and the press reports that all members of both camps were alerted to do what they could for the chap on their side.

This was one of those complicated formulas, with Roeder needing to win with Nixon finishing worse than sixth. When the flag was waved all bets were off, Fred Nix took the lead and kept it, Nixon was second, Roeder was fourth and Nixon was national champion.

If 1967 was a low point, 1968 was the year Dick O'Brien still considers his finest hour.

This was a team! When the KR was a production racer in fact as well as rule, it came in all standard colors. As mentioned, the factory's bikes were Pepper Red, Carroll Resweber rode the Blue Goose, Roeder and the Ohio fast guys' were yellow.

For nearly as long, Harley-Davidson had used orange and black as company colors. The banners had that color scheme, so did the parts boxes.

Well, O'Brien asked himself one day, why not use those colors for the team? And because he was in charge, O'Brien did it.

Next, frames, fairings and riders went to California, to the shop of the Wixom brothers, who designed and built road fairings and who had access to a wind tunnel. They laid out new, smaller fairings, tucked around the Lowboy frame and lower suspension, and then did a cone-shaped rear section for the seat, an area where streamlining is important but usually neglected. The first tries were worse than the old eyeball-designed fairings, but later ones were as slick as they looked. The resulting sleek KRTT was painted orange and black with traces of white (striking, is the word). The team bikes were all done in exactly the same color scheme. (We are about as far from Walter Davidson's "What kind of motorcycle is that?" remark as we're likely to get.)

Not to forget the engine. O'Brien had developed a single-cylinder test engine back in Milwaukee. He had the help of C. R. Axtell and Jerry Branch, as well as Lawwill and Reiman, who both were willing to second-guess the factory's advice, but who also shared their results with the home office.

The KR engine got domed pistons, reshaped combustion chambers and dual carburetors. This last was tricky. The intake ports are, of course, within the cylinders' vee. Even when they are fed by one central

Perhaps the newest KR ever made, this is the machine Rayborn used until the arrival of the XR-750 in mid 1970. Another Andres-built bike, this miler had the Highboy frame, with the crossed tubes on the steering head and a flat, that is, horizontal backbone tube with angled brace above it. That's a KR oil tank but the seat/fender is fiberglass, from an early XR, and the fuel tank is a BSA replica. Brad Andres

chamber, it's a tight turn for the intake tract to make. If the same space is filled with two separate tubes, which is what the intake manifold amounts to, the turn is even tighter. Besides, the Tillotson carb, which has no proper float bowl as such, was never easy to tune.

Against all that, the V-twin has that staggered firing order, which also means a staggered and irregularly spaced intake stroke. This cylinder yanks the air in; then, one and a fraction turns later, the second cylinder does the same but the exhaust valve doesn't close when the intake opens. Things get busy and confused in the chamber common to the two cylinders of a V-twin. But if you have a separate intake tract and carb for each cylinder, well, it may be worth cramping the flow a bit.

And it was. Daytona qualifying began like a dream. The fairings were reckoned to be worth at least 6 mph. The fastest qualifier, determined by runs on the banking alone, was Reiman at 149.080 mph. So give the fairing 6 mph, the dual carbs 4 mph.

On the other side of the battlefield, the BSAs looked terrific and ran 130 mph. The Triumphs did 135, no faster than in the previous year. The Yamaha 350 twin did 147 while the Suzuki 500, also a two-stroke, blew up (and kept people from worrying about two-strokes, to their later surprise).

So, the starter waved his flag and they all thundered into the first turn and to the infield, then out on the banking. When they came out of the last banked turn, pounding down the front straight and past the grandstands, all you could see, wall to wall, was orange and black with a trace of white. Striking! O'Brien nearly burst into tears.

A few laps later, O'Brien nearly burst into tears of a different kind. Reiman had fitted tubeless tires, but the size was different and somehow they wore through the oil tank. Then Markel's engine blew. Lawwill's engine blew. Nix got into a duel with Phil Read—the world-class guy whose only mistake was being in the same time and place as Mike Hailwood—and Nix fell.

"Gas 'er, Calvin!"

Only Cal Rayborn was left of the team. He won, and he lapped the rest of the field. He averaged 101.29 mph for the 200 miles of Daytona.

Except for its tragic ending, this would make a wonderful movie.

Calvin Rayborn was a kid who liked motorcycles. After school, as soon as he was old enough for his license, he zoomed around San Diego working as a shagger, a motorcycle messenger. Eventually, he started racing on pavement but crashed, broke his foot and decided to switch to dirt. He married young and children soon followed. Rayborn worked as a mechanic but didn't have the money to go racing.

Then he attracted the attention of Len Andres, who'd had some good riders since Brad retired.

But they weren't nearly as good as Rayborn. Exactly how this works, nobody can explain but Rayborn was simply a natural, a genius, at road racing. He could (and did) start races next to last and still finish second, right from his first days with an Expert license. Dick O'Brien says the two riders never actually competed because Mike Hailwood was retiring when Rayborn arrived, but he (O'Brien) saw both men in their primes and Rayborn and Hailwood "cornered exactly the same."

That was on pavement. On dirt, also for reasons not understood, Rayborn wasn't as good. There must be something to this because the best road racers of the present day started on dirt: Roberts, Spencer, Lawson, Mamola, Rainey. Name a top road racer, and he began on dirt. Something about abandon—committing the bike to an impossible line then riding it through—can be refined into control on pavement at speeds beyond the possible. Kenny Roberts literally redefined road-racing techniques for the rest of the world when he won his first world title.

But it doesn't work the other way. O'Brien describes this indirectly when he says that Markel didn't have road racing worked out until the day his tank leaked oil on his rear tire. "Of course he was running on his own oil and it was like dirt tracking, it was slippery. What he did was, he slowed down and he gained a couple of seconds a lap.

"He learned something right there. He learned you can't push that damned hard, you just ride up to a point . . . From that day on, he was a road racer."

Rayborn began in control but, although he practiced, he never really got comfortable on dirt. Of his eleven national wins, ten were road races. But that, and the rest of his story, comes later.

Rear suspension

When the KR first appeared, it was a dirt bike and it had the rear wheel and hub solidly mounted to the frame, just like the WR before it and the Indians next to it.

The rigid rear worked best. It was easiest to slide, since there were no bumps or ruts on the carefully groomed horse tracks that motorcycles first raced on. Also, the machine could be lower to the ground because there was no wheel travel to accommodate. The rigid frame was lighter than the springer and, of course, the rules required stock frames so the production racers came that way even if, as was the case with the imports, the American distributors had to have special models for racing.

Some of this applies to motorcycles in general, not just racers. There were bikes with full suspension back in the teens, then suspension was reinvented in the twenties and thirties and fifties, until finally they'd learned how to design swing arms and shock absorbers. And, because of the rules, there was nothing to prevent guys from using KRTTs on flat tracks. There were experiments, like Ralph Berndt using

Girling shocks back in 1961, and the Daytona racers swapping parts back and forth in the beach days.

The prime mover, though, seems to have been Mert Lawwill, who built a springer frame for racing flat track in 1962, before the factory did. His reason, he says, was basic: "It didn't make sense to me, thinking you could corner faster with your wheels off the ground."

But Lawwill wasn't the first. Dick Mann was racing a springer Matchless before that, so was Al Gunter. But they did a *Brer Rabbit*, Lawwill says, and told the Harley riders that they (the Harley guys) were sure lucky getting to use rigid frames while the Matchless racers were forced by the rules to stick with suspension. "They were trying to ward us off," Lawwill says. The rules didn't allow or prohibit rear suspension, Instead, the rules said you had to use an approved or production frame.

Then came short track and Class A racing of a sort, and road races became more like those in Europe and less like a TT on pavement. Road races paid extra points so the factories concentrated there until, in 1963, the AMA Competition Committee voted to retain the Class C frame rules "for road racing," implying that for dirt, frame choice had become free.

The informal arrangements agreed to before the fast-gaining Brits generated a full team allowed the riders and tuners to work out details for themselves. Lawwill's tuner was Jim Belland, who worked during business hours for Dud Perkins, the longest established Harley dealership in the world. They began building frames that used some stock parts, notably the twin tomahawks, but with extended rear rails and all-tube junctions. Lawwill and Belland also worked out a really fast XLR, probably the best yet, that weighed only a few pounds over 300.

The official Harley-Davidson factory-built sprung frame dirt bikes didn't come until 1968 and even then, nobody was forced to ride them.

The technology explains itself. The Highboy frame, used in 1967 and then replaced by the Lowboy for road racing, evolved into the 1968 dirt frame. Because the dirt bike needed more wheel travel, the seat and steering head had to be higher, unwrapped from around the KR engine, just as the Lowboy had been shaped to fit tighter. Here again, though, the factory team used frames built in the shop, for the team. Customers got parts books and found the numbers for the cast-junction frames, heavy or light, long or short, just as before. (This wasn't really a hardship, by the way, because by then there were a flock of special firms—Sonicweld was the best known—building racing frames for use with KR and CR engines, and cheaper than the factory versions.)

Some things change . . .

Finally, we arrive at the change that caused the revolution. By 1968 both the side-valve Harley and the 500 cc BSA and Triumph were outmoded. The customers were voting with their wallets for Sportsters and 650 and 750 cc imports, or for the two-strokes from Japan, which were very fast and increasingly reliable. Roxy Rockwood noted in 1968 that the earlier switch—with the AMA committees flopping back and forth, coming up with a 350 cc limit then voting it out—might have been just down-home politics to the Yanks and Brits, but the Japanese took it seriously. They'd seen a way to get into American racing and had full lines of sporting 350s, just rarin' to go.

Instead, the AMA restructured the Competition Committee. It became the Competition Congress, with delegates from the chartered clubs, professional riders and corporate members.

The Competition Congress convened in late 1968, and blew off the lid. The "time-honored, much maligned" to quote Gordon Jennings, formula of 750 cc side-valve. versus 500 cc overhead valve was replaced with a rule allowing 750 cc displacement with no valve system mentioned. And to keep the Class C spirit, there had to be 200 examples of the engine made, and the engine had to be available for public sale.

This new rule would take effect for the 1969 season, putting paid to the 350 cc limit and the time clauses earlier discussed as ways to ease the burden when Class C was revised, as everybody knew it had to be.

For the reverse kicker, the twist on the twist, this new engine rule would apply for 1969 *only on dirt tracks*. Surprise. Full stop. That must have been buried in the fine print. Nobody would admit to knowing about it in public. Jennings wrote that, privately, the BSA and Triumph camps admitted that they'd done it fast in hopes of getting the jump on Harley-Davidson.

Jennings added that his spies said the KR engine was now up to 60 bhp, and that "it's going to be pretty funny if, after all the moaning and complaining about 'unfair displacement advantages' the old side-valve chuffers beat the under-developed overheads on an inch-for-inch basis."

Then he commented that the action sure did come quickly, and wondered if the AMA ruling body would countermand the congress. It didn't.

. . . Some stay the same

Gary Nixon won the championship for 1968, by being consistent and by having opponents who did themselves in by sheer weight of numbers. Nixon won only three nationals during the season. Markel won five; four half miles and a TT. Rayborn won three of the year's four road races. Fred Nix won six; four miles, one half mile and one short track, the best anybody's ever done without winning the title. Each of the team riders was good at what he did best, but not as good as Nixon, so Nixon edged Nix, 622 to 613.

O'Brien stayed up all winter, so to speak. The team members weren't exactly surprised by the change, in fact they'd been tinkering with a racing 750 for a couple of years. But they'd not been able to get it ready in time, and Harley-Davidson (the company) was sailing in rough seas. The family owners went public to raise capital to take advantage of the motorcycle boom, then went looking for a partner, as in takeover, when not enough came in via the stock offering. After some finagling (which needn't concern us here), AMF was to buy H-D, but there wasn't a pile of loose money for such things as a new racing engine.

O'Brien worked harder, is all. The engine guys devised new exhausts with megaphones, which gave more power at peak and less at other times, meaning the gear ratios had to be even better than perfect. There were different camshafts to work with the pipes, and the dual carbs were refined, or so they hoped.

There were eight team KRTTs built for Daytona 1969. They used the Lowboy frame, the megaphones (which weren't supplied for customers, but there were sketches for making replicas included in the tuning booklet) and the 1968-style fairings. The rear brake was a disc, an Airhart made for a racing car, the front brake was a four-shoe Ceriani, to go with the forks.

O'Brien says these machines, last of their line, and the end of their line, had 58 bhp. Period. That's the best they ever did. Mention press quotes of the day, with horsepower in the 60s, and O'Brien just smiles.

At any rate, Daytona arrived and the Harley-Davidson team was in trouble. The KRs weren't as fast in 1969 as they'd been the previous year. Nix was clocked on the outer banking at 147, Lawwill was in the low 140s, Rayborn and Reiman topped out at 135. And the English, ever so patient, were nowhere. Instead, it was the two-strokes, 350 and 500 cc twins. A Yamaha was on the pole at 150 mph, then a Suzuki

Mert Lawwill's old reliable, the same bike shown earlier, this time posed at a vintage rally, 1986. The front of the frame looked like a Highboy, with over-under top tubes, but it was built by Jim Belland. The tanks came from Lipp Plastics and the seat/fender was from the aftermarket, circa 1968. The bike now has brakes because the present owner used to ride it on the street, which also explains the mufflers. At the rally, Lawwill said he recalled the KR with more than fondness, because when the XR didn't work, while Lawwill was defending his title in 1970, he could always haul out the old KR and depend on it.

at 147.5. At the very last session, holding his breath, came Rayborn, 144.746. Seven of the top ten qualifiers were Yamahas; Rayborn's KR was the only four-stroke in the top ten.

Then it rained. The 200 was delayed for a week and the H-D camp got a chance to find out where the power had gone; it was mostly trouble getting the twin carbs to carburate. Anyway, O'Brien is quick to point out, they weren't *that* slow.

In the actual race, Yvon Duhamel's Yamaha—the fastest bike there—lost its ignition. Ron Grant fell off the fastest Suzuki and Rayborn won, second time in a row, on a machine that, quoting from *Cycle*, "was very definitely not the fastest thing around."

A challenge

Every motorcycle enthusiast must have seen or should be required to see the movie *On Any Sunday*, Bruce Brown's definitive work about motorcycle fun and motorcycle racing.

For those who've seen it, Mert Lawwill needs no introduction. He's portrayed as a businessman and a mechanic and a long-haul driver and a family man and a racer. Lawwill is and was all of those.

He was also the man for this hour.

This was eyeball-to-eyeball stuff. Jennings said, after Daytona, that every one of the top twenty-five bikes at Daytona was illegal and he was prepared to prove it: Harleys had oversize tanks and illegal fairings. Triumphs had special frames and nonproduction five-speed gearboxes, Yamaha had custom heads and cylinders, ditto for Kawasaki and Suzuki and so on. Nobody called his bluff.

The actual confrontation, the head-to-head showdown between the KR and the ohv.750s, took place at Nazareth, Pennsylvania, on a long and fast 1⅛ mile oval. It was the first mile race of the season. Triumph rented the track for early practice and showed up with 650 twins and 750 triples, hottest models on the road and all-legal, as there was no limit

And for the grand finale, Fred Nix's traditional KR, circa 1968. This was probably the newest of the old-style, featuring rigid rear, solo seat with pad and KR tank like Billy Huber's back in 1952. But this frame has had the castings ground small and smooth, and the tubes are thinwall chromoly steel and therefore much lighter than it was supposed to be. Also notice that the last three KRs share Ceriani forks. (None of the tuners minded borrowing ideas or equipment that worked.) Bill Milburn

on the number of cylinders. Nixon began cutting 43 sec. laps on the triple and never bothered to warm up the twin. Then Markel, with a springer rear and a rear disc, and Nix, also on a springer, turned 42s, a 96 mph average.

Nix took an early lead and pulled away, then Nixon reeled him in, then Nix got clear again. At the end of the fifty laps, 56.25 mi., Nix and the side-valve 750-twin had half the straight on Nixon and the ohv 750 triple.

A reporter rushed up to Nixon and asked what happened. Nixon, who never took prisoners and never surrendered, snapped, "I got beat, is what."

And so he did. In that confrontation the various types of machine had mostly equal power-to-weight. Of the twenty qualifiers for the main event, nine were Triumphs, six were Harleys.

In the political arena, Jennings (he was a racer and a reporter who did lots of two-stroke articles as well as Harley stuff, i.e., he was opinionated all right but they were his own opinions) wrote that Harley-Davidson had built a 750 Sportster and asked to be allowed to race it but had been told to first assemble 200 copies. Harley came back wanting assurance the rules wouldn't be changed again. AMA's executive director voted for Harley but was overruled "by a coalition of English and Japanese interests."

On the tracks, Lawwill had good bikes, factory backing and Jim Belland for a tuner. He won three half miles and a TT, and was right in there, finish after finish, so he got the plate, in front of Triumph rider Gene Romero, who didn't win any races, and Rayborn, who won three of the four road races. Again.

What they were like

So the poor, derided flathead went out in a blaze of glory, aided by luck—but then, most winners are. Because of the time span, if for no other reason, the KR must have changed more while staying the same than any other racer or machine, perhaps, in history. It's difficult to say what the KR was like because at the very first, judging from the record, it wasn't much better than the lighter, weaker WR. But, at the other end of its life span, the KR had the legs on the Triumph triple (and that model went on to whip the world's best in production road racing).

There's also the human element. The guys who rode KRs, the Reswebers and Andres and other riders from the days before brakes and suspension, tell us that those days were the best. And perhaps they were, although how can we know?

More useful, maybe, are contemporary accounts. In 1963 *Cycle World* did a track test of a KRTT, the one ridden by top privateer Dick Hammer. With 48 bhp the KRTT would hit 145 mph, revving some beyond its 6800 rpm power peak but not—on this occasion anyway—doing any damage. The rider said the bike felt large and flexible, "decidedly headstrong... but a good rider can get it around just about as fast as anything available."

In 1979, well after the KRTT's day, *Cycle World* did another test, of the 1969 KRTT owned (then) by historian and restorer Steve Wright and ridden in 1969 by Roger Reiman. Or so it was believed; what with the racing shop frames and ownership rules and twenty engines with the same number, it's always best to take historical claims on faith, or with a grain of salt.

The later KRTT was supposed to weigh 355 lb., which sounds right. It was a bear to start, had no power below 4000 rpm and the no-bowl Tillotsons needed persuasion. There was power between 4800 and 6800 rpm. The brakes worked fine except that the rider was used to more brake than this bike had. Beyond that, the rider, a club racer named Pat Egan who was chosen because he was both fast and fearless, and who had no experience on a KR at all, said the KRTT was a racing motorcycle, solid and steady, no problems at all.

What it all comes down to, minus the side-valve versus ohv debates (and politics), is that the KR proves the reverse of the earlier Harley-Davidson slogan: If you can't fix it, better be damned sure it works.

The KR worked.

Chapter 3

Lightweights (1961-86)

In a big country famous for vast distances and big engines, lightweight racing never could become more than a supporting class to the Real Thing, and it never has. The original surprise here is how Harley-Davidson and the AMA—working separately, as you'll see—got into the smaller classes.

Contrary to popular belief, and perhaps working against the American grain, Harley-Davidson decided right after World War II that there was a need for an entry-level motorcycle, a little runabout. H-D (and BSA in England) got rights to manufacture a 125 cc two-stroke from DKW, a German firm better at innovation than politics. The 125 and later the 165 cc verions of this Harley, best known as the Hummer, weren't much of a sales success, nor were they much in competition. Some were used in enduros and desert racing, and a couple were tuned and modified. The high point was a win, in 1956, of the nationally known Jackpine Enduro, but the original Hummer and descendents never actually became racers.

Class A returns

The hero of this story is Walter Davidson, of *that* family.

Walter was a son of a founder, and he was a racing enthusiast, the Davidson to whom the racers went for help. He had a gruff sense of humor and was occasionally tight with a dollar, the sort of man whose friends criticize and then defend just as fiercely if somebody from outside doesn't understand.

Okay. Folklore says that Harley-Davidson owned the AMA, and that the AMA rules were done to favor Harley-Davidson. And as seen earlier, Walter Davidson was very aware of brand names.

Dick O'Brien says when it came down to the crunch, Walter Davidson knew what was important about racing.

"Right after I joined the factory, in 1957 or 1958, they were running the 45s in short track. They were carrying people to the hospital, especially in California and Santa Fe, that track near Chicago. It got worse and worse. Walter Davidson put a proposal through, he forced that proposal through, that short track be changed to 250 cc.

"Some of the competition, they didn't want Harley to drop out. Everybody wants to win but they want to have the other teams out there. They said 'Walter, you guys got anything coming?' He said 'Nope.'

"'Well, you're gonna be out of that picture.' He says 'I don't care. Here's the records of everybody getting killed on the 750s, year after year, by God let's get the boys off those big things.'

"We had the 165s, the two-strokes, but they wouldn't work, not against the Triumphs and the NSUs. In order to get enough motorcycles they had to permit a bunch of things, special frames and suspension and you could take a smaller engine and bore and stroke it to 250, so they referred to it as Class A.

"But that's how Class A and the lightweights came about."

Aermacchi influence

What went around, came around. Harley-Davidson management, with no concern whatsoever for racing, came to the sad conclusion that the little two-strokes weren't going to stem the tide of imports, and that the US plant couldn't compete on price; that is, it couldn't build a small motorcycle in the United States for the same money it could build them elsewhere. However, the excellent Italian company, Aermacchi, had fallen on hard times of its own. Harley-Davidson bought controlling interest in Aermacchi and began importing machines to supplement and then replace the domestic tiddlers, and to fill out the line of models, which then had a gap between the little two-strokes and the Sportster.

The best Aermacchi was a four-stroke 250 cc, with one cylinder sited horizontally in front of the crankcase; not like other engines at the time, but one of several Italian engines designed this way, and by Honda and even, in a sense, by Harley with its fore-and-aft twin of the twenties. Air cooled, of course, four speeds forward, foot shift and hand clutch, kick start, two valves per cylinder, telescopic forks and swing arm rear suspension; a neat and tidy little road machine with full fenders and a generous fuel tank. The first US-market Aermacchi was named Sprint, given the factory code of C—equating with the X-coded Sportster, the F-coded Duo-Glide, the A-coded scooter and the B-coded two-strokes—and was imported in 1961.

The three Cs

Walter Davidson couldn't have predicted this, but he got his own back. The smaller bikes had become popular as transportation and as sport. There were club races for road racers, with displacement classes. There were enduros and scrambles, and there was short-track racing indoors and out.

None of these were terribly sophisticated. The road racers used fairings and had few limits on their engines, but most of the racers were amateurs on small budgets so they, in effect, competed on souped-up road bikes, such as those in Class C. Scrambles were club races too, also on production bikes but with dirt equipment, so that was low tech. And while short track was Class A in the book, that is, full racing engines allowed, there was no place in the world where truly all-out racing engines were built in the 250 class. So what guys did was get Triumph or BSA or Zundapp or Ducati engines and put them into homemade frames.

Starting from scratch was an advantage, and Harley-Aermacchi didn't have to start from scratch because they already were in the racing business.

Not your stereotype Harley-Davidson. This is Don Pink, member of a longtime H-D family and an off-road star before the term was invented. He's racing a CRS Scrambler in an eastern scrambles, a pre-motocross form of competition. Harley-Davidson

Roughly speaking, *Aermacchi* translated into "air machine," by intention the name of a company that began building airplanes. Fighter planes. World War II put paid to that, so Aermacchi switched to scooters and three-wheeled trucks and then sporting motorcycles, with the horizontal cylinder, in 175 and 250 cc. They did so well in Italian national races that they went into Grand Prix racing and became the privateers' favorites in the early sixties because they offered the most bang for the buck. At the same time, because they didn't have that many bucks, the racing engines were highly and skillfully tuned versions of the 250 singles used for road machines on the home market.

And used for the American Aermacchis, the little Harleys, the Sprints... Doesn't it all fall neatly into place?

CRTT

Aermacchi was building Class C racers before it had even heard the words. All Harley-Davidson needed to do was use the road bike (the Sprint), file papers to get recognition for better brakes and different tanks and suspension and so forth, and introduce a road racer based on production pieces.

Model	CRTT
Year	1963
Engine	horizontal single, alloy head and cylinder, ohv
Bore and stroke	2.598x2.835 in.
Displacement	250 cc
Brake horsepower	28.5 (claimed, 1963)
Transmission	4 speeds, 1961-63; 5 speeds, 1962-65
Wheelbase	52 in.
Weight	245 lb. (full tanks)
Wheels	18 in. alloy rims
Tires	2.75x18
Brakes (factory)	8 in. drum f/r

Which is just what the company did. Because the K was the 750 road machine, the KR was the flat tracker and the KRTT was the road racer and TT jumper, the Sprint C became the CRTT in racing form.

There were two perfect places to begin. The Sprint frame was a single, large beam from steering head to swing arm pivot. Not like what Harley had used, but classic in its own way and a sound design, with just the right amount of torsional stiffness and flex.

Second, the Aermacchi engine was already being raced. As sold for the street, the 250 had 19 bhp in

The CRS was the competition version of the road-going 250 cc single (later 350) with the upgraded frame and suspension minus the lights, speedometer, mirrors, front fender and so on. The frame was a massive backbone, with the engine hanging from it and with smaller tubes for the rear suspension and seat. *Harley-Davidson*

sporting form; that is, slightly raised compression ratio and semi-race cam. The racing version had higher compression ratio, hotter cam and a new cylinder head, with larger valves and properly flowed ports. Not only did the private owner not have to polish and flow the head himself, he was advised that the engine had been built, tuned and run before it had been shipped. Fill it with castor-base (bean) oil, warm it well before running hard, but don't break it in, said the owners manual.

CRS

Scrambles racing was just as easy. For most riders there was the Sprint H, a stripped-for-fun model with smaller tank and skimpy fenders. All the factory had to do here was remove the road stuff, upgrade the suspension and install the racing engine (slightly detuned for more low end punch). And there was the CRS, a scrambler that could (and did) also run the professional TT races. As yet another non-coincidence, there appeared lightweight and heavyweight TT rules, with 250 and 900 cc limits in place of the old 45 and 80 cu. in. limits.

CR

This is true, no kidding: The spirit of Class C, seemingly forgotten years before this, was that the ordinary guy could get or build a motorcycle with which to compete, even-up, against the dreadful bullying factories. As you've seen, this ideal suffered speedy erosion under the pressures of pride and sales appeal.

With the C-series racers, Harley-Davidson reintroduced the theory: For professional or club short-track racing there was the CR, named in keeping with the flat-track KR.

Except that the CR was a kit; you didn't get a motorcycle. Instead, you got an engine, a frame, wheels, front suspension, handlebars, tank—everything except the tires and tubes. With the crate of parts came a four-page instruction manual. You pried open the crate, flipped to page one and built your own

Sorry about the mess. The Aermacchi engine was neat and narrow in the front, but the cases weren't. Notice the right side shown here, outboard of the shock. Harley-Davidson

The spirit of Class C. This is how the new guys did it, with parts on hand: The bike is a Sprint H, but with CR engine, both from 1961 or '62. And it worked, because the rider is Gene Romero, who became famous and national champion riding for Triumph, then managed the Honda team to still more titles.

short-tracker at home in your spare time for fun and profit, just like the ads in the home craftsman magazines used to promise.

Remember, when this began, short track was Class A, so the factory needn't have done any of this, or it could have offered the engine only as a ruse to keep the team in the races and the self-supported racers on the sidelines.

Instead, taking the long view, Harley-Davidson offered a way for the privateer pro to stay with the brand and compete, even up, with the teams. (As it turned out, Historian Jerry Hatfield says, most of the true little guys went with other brands in the club ranks because those bikes were easier to get and sponsoring dealers were easier to find. But that doesn't diminish the principle.)

Another point here is that the admission of 250 cc short track to the national championship, Class A or C, set the stage for the unique American format. We now had short track, half mile and mile flat tracks, TTs and road racing. Five types of races, with four different machines needed. Well, you *could* have run a TT bike with fairing for pavement, and without the brakes for flat track, and you *could* have won the national title without racing the annual short-track event, but in a practical way you had to have four machines. And for a few years, 1963 through 1965, you would have needed a 250 for TT and for road racing as well. So the spirit of Class C gives, and the spirit gets taken away from.

Technical highlights

Three new little Class C motorcycles, the CRTT, CRS and CR, had in common their Aermacchi-built engines. And the engines were the important part, the basis for success in the beginning, and the limit (to the road racers, at least) toward the end.

There are several plurals here. There were variations between the short-track engines and the road-race engines, as the needs and demands differed. There was constant development and modification season to season, and there were offshoots dictated by the rules.

The beginning was a sound and basic four-stroke single. The only odd feature in 1961 was that the bore was smaller than the stroke (2.598 in. versus 2.835 in.). Long-stroke engines were thought at one time, presumably when the Aermacchi was designed, to have more torque at low rpm. And before that, engines were taxed on bore size, so it made sense to keep the taxable dimension small and the other dimension big. But Honda's bikes and just about eve-

What's that line? It's a whole new ballgame. Sure enough. This is Santa Fe, the short-track national in 1961. Leading and about to win in the Harley CR's first national race is national champion Carroll Resweber, No. 1. Just behind him but not by much are Bart Markel and Joe Leonard (98), Resweber's teammates. Harley-Davidson

rybody's cars had proved by the sixties that short stroke means safety at high rpm, and big bore means big valves which equals better breathing, and the two add up to power.

For cost reasons, the Aermacchi used one camshaft, in the cases, with pushrods and rocker arms—overhead valves in the usual terminology except that the cylinder was horizontal so the valves weren't overhead. (Never mind, ohv it is.) The engine had two valves: intake above and carb above that, with exhaust below. The cases were wet sump, since the oil was carried in a cavity below the crankcase proper. There were two flywheels pressed together, a built-up crankshaft in normal terms, and construction was of course unit.

The horizontal cylinder was simply part of the equation. Compared with a vertical cylinder, as used by Triumph and Ducati, the lay-down was lower and longer, so the builders had more freedom in how high the engine was in the frame, but less freedom in where, fore versus aft, the engine went within the wheelbase. This can be argued but not settled, as both sides won and lost.

The basic engine was the same from 1961 through 1965. Within that time, though, there were two models. The 1961-62 version had a redline of 8500 rpm, compression ratio of 9.5:1 and a four-speed gearbox. The 1963-65 had redline at 9500, compression ratio of 10.5:1 and a five-speed gearbox, all duly homologated with the AMA. The early engines used battery power for the points and coil, the later ones had self-powered magneto ignition. The early engines had about 25 bhp, the later ones were rated at 28.5.

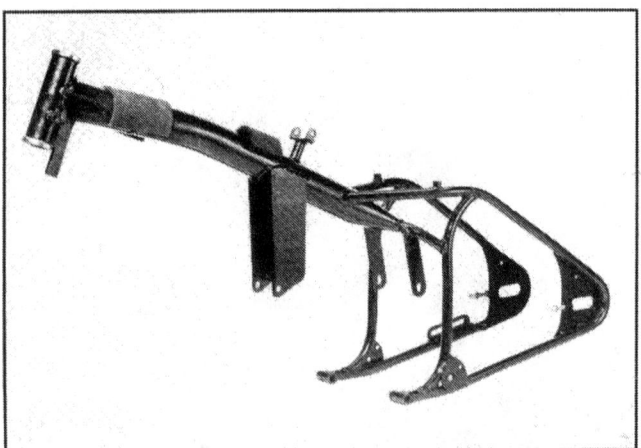

The CR, the short-track machine, was easily made from the Sprint or the CRS. It's the massive backbone frame with the engine hung below it, but for the CR there's a dual framework of small tubes welded to the rear of the backbone and carrying the rear hub, rigidly. The rear of the engine attaches to the lugs at the front of this section, so the engine helped stiffen the structure. This is one of four variations of the CR frame built by the factory team. **Harley-Davidson**

All or most of the above boils down to this, more or less: The tuners were free to tune, and they did. There were at least eight different camshafts listed in the factory's books, never mind how many outside firms were doing parts for the engines. The dirt engines used small carbs, 27 mm versus 30 mm for the road racers, although both sizes came from Dell'Orto, the Italian firm that worked closely with the racers.

The next big jump came in 1966, with a new engine. It was the same overall layout—horizontal cylinder, two valves, pushrods—but the bore and stroke were fashionably oversquare, 2.83x2.40 in. The larger bore allowed a new head with better ports and larger exhaust valve; the improved combustion chamber meant compression ratio could be raised to 12:1; and shorter stroke gave more revs safely, so rated power rose to 35 at 10,000 rpm.

More subtly, the valve rocker arms were relocated on their pivots, changing the ratio; move the pushrod side 1 and the valve side moves 1.59. This

---

Telling the players apart

You can't tell the players without a scorecard, they used to say in baseball. Motorcycle racing has been even worse: The numbers change but the players don't. And the system has changed during the fifty years we're reviewing here.

Way, way back guys either picked their own numbers or they were assigned at every race. That didn't work, so the AMA took the job and assigned numbers, arbitrarily at first. But those who didn't like it, for example Brad Andres, would show up with the numbers they liked better. Then the AMA handed out numbers by finishing order of the previous year but that, too, wasn't popular because who could remember from year to year? The fans wanted to know by gosh when they saw No. 98, it was Joe Leonard.

Finally, in 1971, a balance was reached. The top man in the points race the previous year, the national champion, wore No. 1. Period. Former national champions got the rest of the single-digit numbers; for example, No. 3 for Ricky Graham, and No. 8 for Steve Eklund. The rest of the top ninety-nine experts get the other two-digit numbers and keep them from year to year, so you could tell it was No. 18 Terry Poovy or No. 25 Jon Cornwell or No. 44 Alex Jorgensen. The other Experts, the junior varsity, made do with two-digit numbers and geographical letters.

Thus we first saw Jay Springsteen as 65x, in his rookie Expert year, then as No. 25 because he'd done so well in the points that year, then with No. 1 for three years and with No. 9 ever since.

It's a clearly understood system and it's worked well due to sportsmanship and luck. When Ricky Graham didn't keep the No. 1 plate, he needed a one-digit number. By good luck he was riding for Honda, and Honda's manager was former champ Gene Romero, who officially retired his license. So Graham could ride with Romero's old No. 3.

meant you needed to move the pushrod side less for the same valve lift, so there was less work being done by the camshaft and less stress on the whole valvetrain. In the same way, the carb's float bowl became remote, rubber-cushioned to keep vibration from frothing the fuel. The clutch, which used to be in the primary case and running in oil, was moved to the outside, covered but dry.

There were two sets of gears for the five-speed transmission: close-ratio (5B) and closer still (5A). There was a new frame. It still had a single, large backbone tube and the engine hung from that, but the backbone was larger and more rigid and there were new steel plates to stiffen the swing arm pivot and the engine's location in the frame. The suspension and brakes, 200 mm Oldanis, were supposed to be the same but the team guys said that with the same engine and rider, the new frame lapped Daytona two seconds quicker than the old one did.

The 1966 CRTT was close to the peak, in America at least. There were 1967 and 1968 models in the catalogs and the factory's records show 35 CRTTs built in 1967 (none in 1968) but the only change in specification was the switch to an 18 in. front wheel and the use of Ceriani four-shoe front brake and 21s (twin leading shoe) rear brake, 230 and 200 mm in diameter.

The CRS—the TT and scrambles bike—didn't get the short-stroke/big-bore engine in 1965, but kept the older unit. And the CRS had the four-speed gearbox, a milder camshaft, a 2.4 gal. tank and 7 in. drum brakes front and back.

In 1967 and 1968, the CRS got the improved short-stroke engine with mid-range cam, 10.5:1 compression ratio, the 27 mm carb and a claimed 27 bhp. Factory reports of the day said the CRS weighed 240 lb., which sounds light for a 250 cc dirt bike and heavy

Short track is specialized, and this example, a typcial CR with its Harley engine and over-the-counter parts, is normal: rigid rear end, low frame and engine, no brakes, small tank, steep steering head angle. Huge rear sprocket means the rider will spin the engine 10-12,000 rpm in top gear on the small oval.

Model	CRTT
Year	1966
Engine	horizontal single, alloy head and cylinder, ohv
Bore and stroke	2.83x2.40 in.
Displacement	250 cc
Brake horsepower	35@10,000 rpm
Transmission	5 speeds
Wheelbase	52 in.
Weight	215 lb. (dry)
Wheels	19 in. f/18 in. r
Tires	3.25/3.50
Brakes (factory)	2 1s; drum f/r

if the CRS's smaller brakes and tank are taken into consideration.

From 1966 on the CR engine was listed as the track engine, same as the CRTT except it used a different primary drive ratio, 2.52:1 against the 2.09:1 of the CRTT. This means that the dirt engine spun faster in relation to the clutch and gearbox than the pavement engine did, and that there was more multiplication of torque, more strain on the gears. But on dirt, in shorter races with less shifting, that doesn't matter.

Four different frames were listed for the CR during its production years. They were the same in principle, but there were high and low engines, and there were steel and alloy mounting plates. It would be hard to regulate this because most of the racers who bought the engine didn't buy the rest of the parts, so the engines and frames became jumbled right from the beginning.

As the racers and tuners learned more, suspension became the norm instead of the exception. But the factory never offered a swing-arm frame for short track and Lawwill was the only team rider who really liked rear suspension. The others won with rigids as late as 1970.

Big brothers and cousins

Harley-Davidson and Aermacchi were partners working under two different set of rules. This led to

Short-track rules allowed more freedom than the road-race class did, and there was a brisk business in aftermarket racing equipment. This bike, built by historian/collector Dan Rouitt, is a tribute to Cal Rayborn and wears his national number, 25. It's a tribute, not a replica, because the engine is an early, long-stroke version while the frame, a Trackmaster, was made in 1969 or 1970.

(or perhaps fueled) suspicion and mistrust on both sides later, but it also resulted in some variations on that same, basic engine and frame.

In Italy there was no dirt track like anything in the United States. But there were club and professional road races with lots of displacement classes, and there were world championships (even though this, uh, world stretched all the way across Europe) for 250 and 350 engines.

For the home market, Aermacchi offered a complete line of production racers: The basic name was Ala, or Wing, and the model name, designating displacement, was a color. So there was the Ala Rossa, or Red Wing, 175 cc; the Ala Blu and Ala Verde, blue and green, the road and sports versions of the 250; and later, there was a two-stroke, 125 cc, called the Ala d'Oro, or Gold Wing. Honda paid to use the name, while the two-strokes became the standard motorcycle. But that needn't concern us here.

Because there were displacement classes, Aermacchi in 1963 introduced a 350 cc model. It had a bore and stroke of 2.91x3.15 in. Right, the new engine was undersquare, reportedly because it was cheaper and easier to make the bore as big as it could be made without needed major change to the cases, then increase stroke until the displacement limit was reached.

In the United States this engine came in the ERS, a scrambles and enduro machine. It was rated at 35 bhp at 7500 rpm, with a redline of 8500. It had a compression ratio of 10.5:1, used the 30 mm Dell'Orto and the same size valves as the CRS, along with the same brakes and suspension, wheels, tires and so on.

All the C-series racers were powered by the Aermacchi four-stroke single, called the Sprint in road trim. It was unit construction and had the camshaft in the cases (in the gear case, in Harley terms) with pushrods running through a tunnel to rocker arms under the covers on the head. The racing engine used a magneto, so the chrome cover in the center is a dummy, with the camshaft inboard of where the points used to be. Harley-Davidson

The Sprint engine is a wet sump, carrying the oil at the bottom of the crankcase. Dell'Orto carburetor has a direct shot down the intake tract and the carb's float bowl is remotely mounted, on the frame tube behind the carb body, to cushion it from vibration.

The C-series engine was state of the art in 1961: domed piston with indents for the valves, and hemispherical combustion chamber with valves as big as possible and with the spark plug offset. The cylinder was aluminum with iron sleeve and the hole on the barrel's left was where the pushrods went. Harley-Davidson

The CR racers got new, useful parts, just as the big machines did. This is the long-distance CRTT tank, offered in 1963. *Harley-Davidson*

This was the old rule of cubic inches, for which there is no substitute. The imported Harleys had to contend with some really neat scooters, dirt and dual purpose, from Yamaha, Honda, Hodaka and a host of others, so for play riding and fun, it was easier and less stressful to get the power from a 350 than from a 250. And even though it was legal to race a 350 single against 500 and 750 twins, it wasn't sensible. All the professional races in the Grand National circuit were for big bikes or small bikes but nothing in between. The ERS was a good play bike, but it wasn't a professional racer. For the street and for dual-purpose bikes, the Harley 350 was called the SS-350. It, too, was a good little bike but was outmoded and was taken off the market in 1974.

Meanwhile, back home, the production racer 350 was brought out in 1967, with the five-speed gearbox, stronger frame, a 32 mm Dell'Orto, a four-shoe Ceriani front brake and a host of other improvements. In 1968 the 350s were full works specials, with different bore and stroke—3.031x2.953 in. Back to oversquare. This engine cranked out 42 bhp at the rear wheel, or 50 bhp at the engine sprocket. The cases were so small (the better to get low and narrow) that the flywheel was on the outside. On demanding courses like the Isle of Man, where rider knowledge and sheer handling and ease of maneuver were important, the factory Aermacchi was a match for the Yamaha two-strokes, inch for inch. They were the last four-strokes to make that statement.

The record, and the end

Harley's new models and the new rules couldn't have begun on a better note: The very first Class A short track for 250s and for the national championship took place August 25, 1961. It was won by Carroll Resweber, with Markel and Leonard second and third. (No prize for guessing what they were riding.) The next day a CRTT won a club race in Canada. For three years, 1963 through 1965, there was a lightweight (as in 250) TT race at Peoria as part of the Grand National series. Harleys won all three: Markel twice and Ronnie Rall once. Obviously the CR and CRS were the perfect machines for their class and their events, in their day.

The CRTT was nearly as good—well, it was as good in its own way. The problem was that there was more and better organized, and in some ways superior, opposition in road racing. The Sprint-based CRTT was no slouch, though; it won the lightweight race at Daytona in 1963. And when the AMA—possibly preparing for the proposed 350 cc limit, which never happened—had lightweight national road races in 1964 and 1965, the CRTT won there, too. Dick Hammer won the lightweight at Daytona again in 1964.

But later that year Dick Mann won the 250 national at Nelson Ledges, Ohio, for Yamaha. This was the first national win for Yamaha and the first for a two-stroke, although the 350 Yamaha twins had been getting faster and faster and closer and closer to the 500s and 750s in the big national races.

For Daytona's lightweight race in 1965, it was all Yamaha, taking the first four places; then came two CRTTs, then four more Yamahas. That team went on to win every major road race it entered.

Late in the C-series engine's life, the team tried all sorts of techniques to stave off the Yamaha two-strokes. One of the changes was to a short connecting rod, with cut-down cylinder and a highly modified piston, all shown here on the left, compared with the stock short-stroke parts on the right. The later engines did work. The best one ran in and won the Novice (250 cc) race at Daytona, 1968, ridden by Don Hollingsworth against a fleet of Yamaha twins. That engine was an Aermacchi two-valve; the H-D team wanted to run a four-valve. Anyway, the four-stroke single was bound to lose to the two-stroke twin eventually.

For 1966 the CRTT got the new engine with 35 bhp, a gain of six. But the other factories were working just as hard, and the best Harley at Daytona was Reiman. In fifth.

In 1967 all the stops were pulled out. Lawwill had a short-rod engine with every modification known, and the latest fairing and those huge brakes, but the Yamahas and Kawasakis walked away with it. After the Laconia meet, with the same results, the Harley Sprints were withdrawn from competition and the team riders, most notably Rayborn, who was the only team member really there to road race, were released to ride other lightweights if they wished.

What happened can be measured on several levels. At the most basic, full-race two-cylinder two-strokes had more power than their four-stroke, one-cylinder competition. They had more power because smart and determined men had learned how to get power out of the design. Simple as that.

Dick O'Brien had a guest at that 1967 Daytona debacle; he'd arranged to have an Italian tuner there. Up front, this was so the man from the other factory could help. In fact, O'Brien had him there to learn and observe. The four-strokes were still winning in Europe then, and the Europeans thought (or so O'Brien believes) that it was because they knew how to tune engines and the Americans didn't.

O'Brien said, "They built some special jobs and they came over with the machines. The first day Ezio was very, very down. He got some pictures of those two-strokes flying. We could not stay with 'em, no way, shape or form.

"We decided on an overhead cam, four-valve engine. Got an okay to do it because Harley owned half

The CRTT had a glorious career, when it was like against like. This is Dick Hammer, factory-backed to a win in the lightweight class at Daytona in 1963, with Norris Rancourt, Parilla single, in second place. Daytona Speedway

of the company. We wanted four valves; Mr. Bianchi, the chief engineer over there didn't feel it was necessary but he said okay. They made some aluminum castings and sent them over here for us to develop the ports and the combustion chambers for the four-valve, which we did, spent no end of hours on it. Axtell worked with us on those heads because at the time we didn't have a flow bench at the factory.

"So, instead of waiting to get the four-valve, they went ahead with the two-valve. All they succeeded in doing was increasing the rpm. Peak horsepower was at 11,000 rpm, between thirty-six and thirty-eight horsepower at the rear wheel. Redline and peak were at the same point with the rocker arm engine but the new engine could go to 12,000 . . . still with the same horsepower.

"As soon as they saw there was no horsepower increase they killed the program. I argued with 'em, said if we can go four-valve we can pick up quite a few horsepower but they refused to do it."

The factory, Aermacchi that is, did build a few examples. They did well in some venues, mostly where the Yamahas weren't. Yamaha, of course, began racing seriously in Europe, along with the other Japanese firms, and two-strokes became the rule (play on words is unavoidable).

Which didn't mean Harley-Davidson/Aermacchi had to get out of racing, however.

RR-250

The Harley-Davidson factory team was officially out of the lightweight road races. That didn't mean no more Sprints. Lightweight short track remains as part of the national championship today, and the little four-strokes were part of the team until the motocross engines arrived (more about that later).

Nor were the Aermacchi racers slow to learn that two-strokes could run, and that Aermacchi could make them run with their rivals. While making a splendid gesture in the form of a dohc 350 single that had an astonishing claimed 60 bhp and won the 1971 European hillclimb championship, Aermacchi drew upon its two-stroke experience and built a 250 cc two-stroke twin, a pure road racer, and followed with a 350 cc version.

The team had attracted a fine talent, Walter Villa. The bike began with an air-cooled engine, then added water jackets when the power increased. Villa won the world 250 championship in 1974, 1975 and 1976 and the 350 cc title in 1977. That 1974 series victory, by the way, was Harley-Davidson's first world title.

Back in the United States, things had changed a lot. As detailed later, the rules had been revised and loosened. In brief, twenty-five examples constituted production and an engine was the same as a motorcycle. Although there was no longer the threat of a 350 cc limit, there was an active lightweight, 250 limit, class. So Harley-Davidson did the sensible thing and homologated Aermacchi's racer as the Harley-Davidson RR-250, offered for sale to the public and thus eligible for AMA Class C professional racing.

That was virtually the only link between the RR-250 and production. The frame was a duplex cradle of twin steel tubes, shaped like a diamond laid on one

CRTT (and the KRTT) began life as Evolution models. Alongside his KRTT, this is Dick Hammer's 1963 Daytona lightweight winner, as tested by *Cycle World*. The early

CRTT was pretty much a hopped-up Sprint engine in a Sprint frame, with big brakes and fairing and larger gas tank added. *Cycle World*

side above and below the engine and with a double triangle extending back to the seat support and upper shock mount. The customer bikes came with Ceriani forks and four-shoe drum brakes, although the factory's own racers used dual discs in front, while keeping the Ceriani forks. Rear brake on both versions was a twin leading-shoe drum.

The engine was an inline twin, inclined forward 15 deg. from vertical and with classical oversquare dimensions of 2.20x1.96 in., more often expressed as 56x50 mm. (The other classic dimension for a 250 cc twin is 54x54 mm.) The dry clutch and gear-driven water pump were on the right, with gear primary drive from the crankcase to a six-speed gearbox. The Italian versions used Dell'Orto carbs but the RR-250s shipped to the United States came without carbs and were fitted, usually, with 34 mm Mikunis; these were readily available, with plenty of parts on hand, and it was generally agreed they made for easier starting.

Helpful note here, the factory's booklet lists two compression ratios. The European method has usually been to compute compression ratios for two-strokes the same way it's done for four-strokes; that is,

Model	RR-250
Year	1974-76
Engine	water-cooled two-stroke twin
Bore and stroke	2.20x1.96 in.
Displacement	246 cc
Brake horsepower	53 (claimed)
Transmission	6 speeds
Wheelbase	55 in.
Weight	240 lb. (dry)
Wheels	18 in.
Tires	3.25 f/3.50 r
Brakes (factory)	41s Ceriani f, 21s r

measure the difference between the displacement of the combustion chamber and the chamber and cylinder.

The Japanese use—and have popularized—an equally logical but different idea. Because compression doesn't actually begin until the cylinder is closed off, until the ports of a two-stroke are covered, they calculate the cylinder's swept volume at the moment

The 1964 team CRTT had Ceriani forks and two leading shoe front brakes, while the customers had to make do with the standard Sprint front end. Everybody got the smaller and more racelike fairing, though.

the ports are covered by the piston and compute compression ratio from there.

This means that the RR-250 had a compression ratio of 12.8:1 in the West, 6.6:1 in the East. It also means radial compression and radical port timing and yes, the engine didn't begin to work until 6000 rpm, with claimed peak power of 53 bhp at 11,800 rpm and a redline of 12,000 rpm.

Cycle World tested Gary Scott's RR-250 late in 1975, the year Scott won the national championship. Scott was a willing and diligent road racer, if not the most naturally gifted. The magazine staff found the RR-250 to be exactly what they'd expected, a full-race machine with all the speed and brakes one could wish for, especially with the slick tires, just then introduced by Goodyear. The expectation was that Scott would do in the United States what Villa had done and was doing in Europe.

It didn't happen. Scott won the lightweight class at the Louden national meet in 1974, first race for the new bike, but that was the RR-250's peak.

Why it didn't happen begins with Kenny Roberts. Scott had Roberts to contend with, while Villa only had the rest of, what the Europeans call, the world. And when Roberts went to Europe, he ripped the entire GP circus into little pieces, which they couldn't believe but Scott could have told them about.

Yamaha took the whole thing seriously; at home, in Europe and in the United States. Harley-Davidson wasn't selling two-stroke twins, Aermacchi wasn't selling in the United States . . . nobody surrendered or walked away, exactly, but neither did they stay up all

The short-stroke engine, shown here in a 1966 CRTT, was, naturally, shorter and wider. Frame, engine hangers, seat and tank were revised for that year.

night or fund research from other places. So Roberts waxed Scott later in 1974, at Laguna Seca, then Ron Pierce was second to Roberts at Ontario. Scott was second to Roberts in the 250 class at Daytona in 1975, while Jay Springsteen (*much more* about him later) was third in the Novice race, 250 limit, at Laguna Seca.

For the national meeting at Ontario that year, the factory imported two outsiders, Villa and Michele Rougerie, to ride with (or block for, or instruct) Scott. But Ontario was flat, featureless and hot. Roberts did his usual, and the guest stars faded. Scott was fourth in the lightweight Expert race, Springsteen was fourth in the Novice 250 race. In 1976 Springsteen was the victim of mechanical failure in the lightweight class at Daytona. (AMA rules then had every rider regarded as a novice in every class, until proven otherwise. So Springsteen, well on his way to his first national championship then, was still ranked as a novice in road racing while ruling the dirt. Rules are odd.)

But the RR-250 had had its brief day.

That sounds grimmer than it was, unless you can imagine not winning as anything except a disaster. But a good part of the RR-250's unimpressive results was caused by circumstance. Brent Thompson, who was on the team and worked on the 250s, remembers that the engines used in the United States weren't exactly what the Italian branch used, and that some of the facts and techniques (as in oil mixture and torque settings, even the order in which one torqued the various bolts) were not clear in the beginning. Then there were the different carburetors for the American RR-250s. And there were time problems:

Team fairing, here on George Roeder's CRTT at Daytona in 1966, provided fuller coverage than the first model had. The two-shoe brakes (notice there's no levers on the right side) were a small version of the Oldani used on the KRTT. Harley-Davidson

The lightweight tuners didn't have the time they needed on the dynamometer to work out the bugs.

At the end, the RR-250 was better than it had been when it won. Thing was, the other chaps hadn't been sitting still. One race report from that era says that Scott's was the fastest "stock" 250, meaning Roberts' Yamaha was special. Which of course it was. It was also legal, Class C having been left a long way back.

The team did what could be done, even bringing over one experimental RR-250 with single rear shock. Routine now, but radical in 1976, when dreaded Yamaha was making the design popular.

Harley-Davidson had moved into lightweights, but not because the small machines fit the factory's image; they didn't, not by yards and yards. Harley was in the road-racing 250 class because the multicylinder machines, the fours and threes from Yamaha, Suzuki and Kawasaki, had taken the big class away from the twins and triples from BSA and Triumph as well as Harley, while the English factories had also gone out of business.

There was more raw material in the Aermacchi shops. Harley-Davidson could try what Yamaha had already done, take the 350 two-stroke twin and make it into a 750 cc four. Aermacchi already had a 370 twin.

RR-500

Aermacchi also was at work on a 500 twin, an incredible exercise. The built-up four seemed too complex and too far from what Harley-Davidson was doing the rest of the time, so the big twin was homologated and introduced as a production racer.

This was, at best, a slick way to get the machine into American racing. The rules had been modified to allow racing-only engines. Not in so many words, but all the team needed was one motorcycle and proof that at least twenty-four engines—just engines, not frames or suspension or spare parts—were available for sale in the United States.

The RR-500 began life as big brother to the RR-250: same type and general design of frame, with a water-cooled two-stroke twin, six forward speeds and so on. But it was more complicated as well as bigger.

The specifications tell only part of the story. The most visible and notable feature was the rank of carburetors: four—count 'em, four—34 mm Dell'Ortos lined up aft of the cylinders. The carbs fed reed valves and then the crankcase, all normal racing practice of the day, except that because large two-stroke twins were rare (indeed it wasn't too long before this that everybody knew you couldn't build a two-stroke this big because it couldn't dissipate heat fast enough), nobody made carburetors large enough for such an engine, so they needed to double up. And they did look impressive.

The compression ratio is listed twice in the RR-500's specification book: it was 5.07:1 from the closing

Dual disc front brakes were homologated for the CRTT in 1966 but not used until later and then not in all the races. They were experimental and didn't always work, while everybody understood how to get the most out of drum brakes. Harley-Davidson

One of the last production four-strokes from Aermacchi, this is the 1970 customer machine, which did well in Europe in the 350 class, then as a clubman's bike, and after that in vintage races. Note the Ceriani forks and the four-shoe front brake. *Cycle World*

George Roeder on a busman's holiday: The machine is one of two streamliners built by James "Stormy" Mangham, an American Airlines pilot from Texas, and used by Harley-Davidson on the Bonneville Salt Flats. Roger Reiman got 149.78 mph out of this streamliner with the long-stroke Sprint engine in 1964, then Roeder and the short-stroke Sprint were clocked at 176.817 mph in 1966. Harley-Davidson

Model	RR-500
Year	1975
Engine	water-cooled two-stroke twin
Bore and stroke	2.84x2.36 in.
Displacement	29.8 cu. in. (488.3 cc)
Brake horsepower	105 (claimed)
Transmission	6 speeds
Wheelbase	55.5 in.
Weight	265 lb. (dry)
Wheels	18 in.
Tires	3.25/3.75 or 5.00
Brakes (factory)	dual 11 in. disc, f/8 in. drum r

The two-stroke racing engine was made, almost literally, by joining two 125 cc singles. The 250 had two sets of flywheels, each running in two main bearings and joined by the clamp shown between them. The left crank drove the alternator for the ignition, the right one had the gear for the primary drive. *Cycle World*

of the ports to top dead center, and 12.1:1 from the bottom of the stroke to the top of the stroke, as detailed in the RR-250 description.

The book shows two horsepower ratings as well. The power is given as 89 bhp, measured at the rear wheel, at 10,000 rpm. In Italy, according to Ted Pratt, who owns the RR-500 shown here, the rating was 105 bhp at the engine sprocket.

Both figures are impressive: In 1975 educated guessers said a good flat-track Norton had 75 bhp, a Yamaha 750 twin had 70 and the newly improved Harley XR had 85. Some of this might have been bluff, but the ratio probably would hold up under scrutiny.

The RR-500 also had sizable brakes, and because big tires and slicks for pavement racing were becoming the order of the day, there shouldn't have been any lack of traction. The RR-500 had six inches of ground clearance, five inches of front wheel travel and three inches of rear wheel travel; not a lot of travel by modern standards but normal then.

The factory's data bank doesn't hold much back. Steering head angle was 28 deg. There was 3.3 in. of trail, the distance between the front tire's contact with the ground and a line drawn through the steering axis—a factor in keeping the bike straight at speed. The tank held 6.3 gal. of pre-mix and claimed to top speed with stock engine and factory gearing was 175 mph.

All this sounds impressive and promising. Because the 500 cc class in Europe was the premier class,

Aermacchi switched to two-strokes, two cylinders just in time. This was the customer version, made in 1972 in 250 and 350 cc form, with air cooling. Ceriani front end and brakes carried over from the four-stroke 350. *Cycle World*

The Aermacchi racer's frame was small tubes formed into a duplex cradle for the engine, backbone and shock mounts; a space frame, in car terms. Cooling the power meant huge fins. This engine was unusual in that crankcase and gearbox were separate assemblies, like the old-style motorcycle engines and like Harley's big twin. (Beyond that, no two engines ever had less in common.)

Aermacchi in Italy should have been just as keen as Harley-Davidson in the United States, and it probably was. Nor was there a lack of experimentation. One RR-500 was made with the rear brake, a disc, on the countershaft sprocket where it could be smaller and lighter for the same effect, and where in theory the braking forces could be fed directly from the frame to the wheel without the suspension knowing. (This may have happened, but it didn't work well enough to be used in any races. The machine was later crushed by the racing department.)

There were high hopes for other reasons. Prior to the OPEC blackmail, Europe was a good Harley-Davidson market. H-D and Aermacchi had plans to expand their joint sales efforts there.

The H-D team went to Europe in 1973. It was in some ways a promotional tour, and in other ways a test.

The team's top road racer was Gary Scott, who remembers that the Aermacchi shop was glad to meet him.

"Those guys, Walter Villa and the team, they were used to 250s and 350s and they were afraid of the 500. They weren't used to all the torque and they rode the 500 like a 250, revving it," Scott says.

Scott was dealing every day with the alloy XR-750 as well as the RR-250, so the 500 seemed to him like a nice, mild compromise. They went to Monza for some testing and Scott broke the short-course record.

The Aermacchi crew, he says, "were all in love with me."

Then they went to Austria, the Salzburghring.

O'Brien says, "The Italians were tickled with Gary because when they were practicing it was in the rain. Against all your hot shots, your Grand Prix riders, Gary was on the leader board with the third fastest lap."

What happened on race day is hard to credit, considering that these were two of the world's top teams, combined.

O'Brien says, "Where the failure came was, it looked like it was going to rain, and felt like it was going to rain and it was a matter of, which tire do we put on? We didn't have extra wheels [That's the part that's hard to credit] so we went over and got all-round Dunlops, a combination tire. Gary elected to

Gary Scott introduced the RR-250 to American crowds and won the bike's debut race, giving hope that would never be fulfilled. Harley-Davidson

Line of descent is clear here: frame, suspension, brakes, seat, and tank; obviously from the air-cooled 250 but now with water cooling and more precise control of heat and heat-caused expansion. Harley-Davidson

Walter Villa won Harley/Aermacchi the parternship's first and only world championship. Harley-Davidson

leave the rain tires on. That was a mistake because he went down that front chute the first time and tried to stop as if it was a dry tire, he just kept sailing, right into the haybales and took himself out the first lap. He would have finished well up there."

Scott says they only had the one set of wheels, but he remembers simply trying to keep up with the other guys and spinning out on the third or fourth lap.

But the result, the important part, was an impressive if-only. They were encouraged enough to go through the homologation ritual and bring the RR-500 into the United States as a certified racing bike. They took it to the national meet at Laguna Seca, 1975. Scott was near the front, but not leading, when a hose clamp loosened and the coolant escaped and he retired.

Scott survived and went on to win the No. 1 plate, but he did it on the dirt. The RR-500 never appeared as a Harley-Davidson again. Some of the engines went to the Italian tuner, Bimota, and were raced in Bimota chassis, while two or three RR-500s survived in private hands although both of the two known to be in the United States came here from Europe and not from Milwaukee.

Why? In general, Scott says, the RR-500 he rode at Laguna Seca wasn't competitive. Further, "It wasn't the same machine" in that it hadn't been prepared and tuned and revised and honed into shape the way

The AMA rules called for only twenty-five examples when the RR-250 came out, but it was shown in the catalog as neat and tidy as it looks here, and some were sold to private teams. *Harley-Davidson*

Walter Villa, Harley-Davidson's—well, make that Aermacchi's—man in the winner's circle. At the height of the partnership's success, the early seventies, the American and Italian teams traveled back and forth and enjoyed each other's company, so to speak. Harley-Davidson

the Aermacchi 500 had been. Nor did Harley-Davidson have the personnel or money to make the RR-500 into a really competitive bike.

O'Brien agrees: "The one Gary rode over there was very fast compared to the one we had here. It needed more development work done... We would have had to have a rule change because of the four carburetors and I don't think any of us felt it was going to be competitive."

"It wasn't a Harley-looking motor," I said.

"No, but if it had been competitive, they would have gone ahead anyway. I don't think any of us felt it could keep up with a four-cylinder 750, and we would have had to enlarge our department and that, too, was a case where they didn't want to spend more money... and that's why we didn't go any further with it."

Circumstances alter racing departments. In retrospect, although it does seem that the bike got a lot of work and money early and not enough later, the decision was a sound one. The rules had been changed again, and although Suzuki did manage one win with a 500 cc two-stroke twin in the national class, the multis, the 750s and the two-strokes were too strong a combination, and Harley-Davidson retired from road racing.

The RR-500 was, no surprise, a larger and heavier version of the RR-250. This is one of the homologation photos taken of an early example. It has a two-shoe front brake and . . .

Wall-to-wall carburetors were there because it was easier and more efficient to double the number of carbs when the displacement doubled, than to find a pair of carbs twice as big. Still, it does look incredible. Ted Pratt

MX-250

Along about the time the overhead 500s began running with the side-valve 750s, say the late sixties, America was introduced to motocross. The date is

The RR-500 had two separate cylinders that bolted onto the crankcase. At the rear of the barrel were two huge ports, with reed valves (not installed here). These valves were one-way: they let the fresh charge into the crankcase and then closed so the charge could be compressed. Ted Pratt

... a disc rear brake, on the engine sprocket instead of the rear hub. This was smaller and lighter than a brake on the wheel would have been, and it took the stress of braking out of the rear suspension, while feeding it into the drive chain. This particular RR-500, the only one built with this brake, was crushed by the factory when it was surplus. Harley-Davidson

Model	MX-250
Year	1977-78
Engine	two-stroke single
Bore and stroke	2.84x2.35 in.
Displacement	15 cu. in. (250 cc)
Brake horsepower	32 (claimed)
Transmission	5 speeds
Wheelbase	58 in.
Weight	240 lb. (dry)
Wheels	21 in. f/18 in. r
Tires	3.00/4.50
Brakes (factory)	5½ in. drum f/r

vague here, because although some people knew that the Europeans raced across unprepared terrain around long, winding, rough and wet courses, not many people watched or rode until a promoter for Husqvarna, a Swedish brand, started putting on professional races and sales displays.

Naturally, people loved the whole idea and naturally the AMA was slow to even admit there was such racing, never mind sanctioning motocross races or having championships or—which would have been best—adapting the new class for the Grand National series.

Well. Motocross boomed. Two-strokes did the same, and Harley's Aermacchi-based four-stroke singles, the ones used to build and certify the CR, CRTT and CRS, went off the market. They didn't sell, not against the Yamaha DT-1, Honda SL-350 and later Elsinores.

To Harley's and Aermacchi's credit, they switched in order to fight back. The four-stroke singles were discontinued and the line of two-strokes expanded. The mainstay was a 250 cc single, built as either a road bike, with low fenders, or as a dual-purpose bike, with high fenders and semi-knob tires but with street gear.

These models, the SS-250 and SX-250, were designed for the US market. Enduro-style motorcycles didn't come into vogue in Europe until they captured the public's fancy at the Paris-Dakar (Africa) Rally some ten years later.

The 250 engine was used as the basis for designing a motocross machine. The AMA had different rules for motocross, more like pure racing or Class A than production. But because the factory hoped to produce and sell the bikes, they were laid out for mass production.

With the design drawn and the parts ordered, the factory began a two-part program. The first part was to put together a racing team, to compete in the AMA's national series as well as selected and related events. The second part was to make 100 prototypes that would be loaned out to racers and racing-oriented dealerships for use, display and evaluation.

The motocross team was backed by the road racing team's two-stroke section, which shifted over from the old project to the new one. Main force was Don Habermill, aided by John Ingraham, Mark Tuttle, Sr., and Brent Thompson. The original riding team

The RR-500 was bulkier than its little brother but otherwise followed the same design lines and systems. This is one of the last made and has twin disc brakes in front and a big drum brake, on the hub, in back. Ted Pratt

Experimental rear suspension, shown here on the MX-250 preserved in the racing department's tiny museum, looked and worked like upside-down front forks—sliders and stanchion tubes, with enclosed springs and damping rods and valves. This system displayed no advantage.

was Marty Tripes, Rex Staten and Rich Eierstadt. They left at the end of 1977, and one rider, Mickey Boone, was hired for 1978. He quit after the first race in the winter series, and was replaced by Don Kuldowski, who promptly won the first meet he rode in for the factory.

The program was properly organized and conducted. The prototype frames were made in Milwaukee, while the team's frames came from C&J, one of the best in the business. The engines were prepped and modified in the United States, for the conditions. There was no false pride. The team bikes had Yamaha wheels and Kayaba forks, from you-know-where, but used by Harley because they worked.

The prototype MX-250 had one distinguishing feature. The rear suspension units looked like the sliders and stanchion tubes seen on front forks and they worked the same way, that is, they had long, narrow springs and lots of oil inside the tubes instead of the springs outside and small amounts of oil inside. This made sense in theory, as the more fluid available to control each inch of wheel movement, the more precise the control can be. Or so the book says. The rear suspension didn't work any better than conventional dual shocks did, so dual shocks were fitted when the actual production bike was introduced.

Failure in this case not being instructive, all one needs to say is, the MX-250 didn't win. The machines were top ten and so were the riders. That's about all they did in national competition during 1977 and 1978: fifth here, seventh there, an overall win someplace else.

Rex Staten, right, was the most successful of Harley's motocross team riders, but none of them ever scored a major championship. The MX-250 was a conventional machine in its production form shown here. Harley-Davidson

There wasn't anything *wrong* with the MX-250. *Cycle World* tested a pilot model late in 1977 and found the bike was as fast, as agile, had as much wheel travel and top speed as anything in the class. The engine was pipey, needed some attention coming off corners and the suspension benefited from careful preparation, but there were no glaring glitches.

What was wrong was that OK wasn't enough. The MX-250, along with all the European motocrossers, was falling behind because the Big Four from Japan were ready to commit as much time and money as it took to develop the motocross market. That meant winning and fighting each other, into the ground if need be.

So at the end of 1978, the motocross program was dropped.

Nothing personal there, either. Aermacchi in Europe did the same thing with the road racing team there, and for the same reason: It couldn't afford to compete against the Japanese factories, and even if it did, the rewards didn't justify the effort. (One former Harley dealer in Europe, that is, one man who used to have a Harley agency there, says he suspects that Aermacchi's racing department took so much time and emotional investment that it drained the rest of the factory of all the energy that should have gone into making better street machines. And it could have happened just that way.)

MX Short Trackers
Where each cloud will have its silver lining, nobody can predict. In this case, because Harley-Davidson did bring in several hundred MX engines and a few thousand SS and SX models, there was no problem establishing that the 250 cc two-stroke single was a production engine and thus eligible for Class C, Grand National racing.

When last discussed, short track had been reinvented as a lightweight class for racing engines, then shifted into a form of Class C, with production engines and some choice of where the other components came from.

That was good for the sport, and because a short track is a circle or oval of a quarter mile or less in circumference, it's relatively cheap and easy to build a track and put on races. Short track became locally popular, in the midwest and in northern California especially.

But Harley-Davidson, the guys who forced lightweight short track into being, had no place. The Sprint-based CR won its last national short track, Bart Markel up, at Santa Fe in 1970.

However. The 250 two-stroke was a wonderful base for a short-track engine. And the Class C rules by

End of the drought, as Jay Springsteen takes a victory lap at the Houston Astrodome in 1977. This team bike had a tuned MX-250 engine, Trackmaster frame, motocross tank, accessory fender and seat, a disc rear brake and Ceriani forks. *Cycle World*

When the MX-250 was first shown to the press in 1975, at the Milwaukee plant, the rear suspension had been made with tubes and sliders that began life as front fork parts. *Cycle World*

1977 said that an engine equaled a motorcycle. So it was easy and cheap and logical—and successful—for the racing team to collect some engines, tune them for track racing, install them in short-track frames and get back into that class.

Which is just what happened. Jay Springsteen won the short-track race at Houston, 1977, and ended the drought.

The one hitch here is that although the factory team was racing in the class, and with machines that could pass muster as Harley-Davidson products, in fact there wasn't anything that could honestly be called a production short-track bike.

The team built its own. Not only that, each team inside the team—that is, each rider and tuner partnership—was free to get the job done as it chose. So Springsteen used an MX-250 engine in a Trackmaster frame in 1978, then had a Champion frame after that. Scott Parker got his first short-track win at the Santa Fe National in 1980, riding Springsteen's bike, while Randy Goss used a Knight frame for his factory bike.

The factory meanwhile did list, somewhere in the literature, the engine used in Class C short track. The designation was MX-250 Short Tracker and it was the engine, by itself, although it was unit construction so you got the gearbox along with the cases and cylinder, barrel and so forth.

With the engine, the customer could get a parts list and a choice of recognized, if not quite approved, frame, suspension, tire and wheel builders. There was no approval as such; in fact, the pamphlet says most explicitly that what you are buying is an engine. It isn't a motorcycle and it isn't a kit. Just an engine. There was no official explanation for this, which seems odd because in the CR days the factory would help. Pre-

Team rider Ted Boody, at Houston 1977, used the same basic package but with a different seat and rear fender, and a Boss frame. The MX-based short-trackers were used until the rules changed and allowed larger four-stroke engines. Harley-Davidson.

sumably, the Voice of the Litigant was being heard in the land.

And being fair, this wasn't any trouble. National racing was booming. For instance 60,000 fans showed up for the season's opening twin bill at the Houston Astrodome in the early seventies, and there was a lot of club racing. So C&J, Trackmaster, Knight, Champion, Panther and others made frames, while Bates and Wixom and Lipp offered plastic and fiberglass tanks, seats and so on. And there were shortcuts, such as the Harley factory telling MX-250 short-track buyers that the exhaust pipe from the Honda 250 motocrosser would work fine. And it did.

But there was no production model, no one standard, so there's no official specification table for the team short trackers.

Short track is its own form of racing, though. Because the tracks are the smoothest of the dirt tracks, there's no need for lots of suspension travel so the bike can be lower to the ground than, say, a TT bike can be. Because short track is almost all turning, getting around corners is more important than keeping the bike straight, so usually a short tracker will have a short wheelbase, 53 in. or less, and a steeply raked steering head. There's less stress because there's less power, and races are shorter so less fuel is carried. The 250 short trackers weighed not much over 200 lb., the 360s a few more pounds than that.

Power isn't as important in short track as it is in the other types of Class C races, but power characteristics are. The engine must have lots of mid-range punch and be completely controllable. No wild surges into the power band. So the 250s had about 30 bhp, about the same as the engine developed in motocross form.

There were some variations. The Aermacchi cylinder heads used to crack under the heat, and the rules required the engines to retain their major castings. But the H-D team machinists worked out a way to first, design newer and stronger lower halves for the head; second, to machine away the weaker portions of the stock head; and, third, weld the two parts,

Harley-Davidson 500-R, also known as the 500-HDR and built as a 560 and a 600, was a Harley by courtesy and by the rules only: Harley-Davidson arranged to buy a version of the Austrian Rotax engine, had it homologated, and then offered for sale to and through H-D dealers. Everything except the engine was sold by and through other people. *Cycle World*

new lower half and old upper half, fins and all, into one bulletproof piece.

Short track is also different in that the riding demands its own style, or even talent. The shorter course and the constant turning can be confusing. Along with other requirements—balance, throttle control and lightning reflexes—a good short tracker, like a star basketball player, must have what John McPhee called *a sense of where you are*, that is, constant awareness of your position and that of the other guys, all the time, at speed and while you're doing other things. (This sense of position and, one assumes, will to win, are probably the only characteristics motorcycle racers and basketball players have in common.)

For all its popularity as a club event, short tracks were local and their venues generally were small, so they didn't have vast grandstands, which meant they may not have had enough seats to allow a large enough crowd to pay for a national race. So, there are usually no more than two short-track nationals every year, which is good for having different kinds of races, but hard on the guy who needs a separate motorcycle for only two races each year (unless he lives in a racing hotbed with local tracks and a pro schedule all during the season).

It also means that the lapse, 1970 to 1977, during which Harley-Davidson didn't win any short-track nationals, wasn't as awful as it would have been if the racing was on half-mile tracks.

Even so, during the 1977-83 seasons, inclusive, there were fifteen short track nationals. Harleys won four, Yamaha six, Honda two and Can-Am three; so Harley-Davidson was back in the picture.

Then the AMA changed the picture. There was concern in the AMA about having to have several different machines, and there was a shortage of new pro riders, the Juniors. They were limited to 500 cc four-strokes. So—also considering that the 750s were too big and heavy for some of the TTs—the rules were changed and 500 cc four-strokes were the rule for short track as well as the custom for the tighter TTs, beginning with the Camel Pro series of 1984.

500-R (HDR)

The only drawback for Harley-Davidson was a total lack of suitable equipment. The factory hadn't

This is the first Harley-Davidson 500-R, with the decals not yet affixed, parked outside Ron Wood's shop. It's set for TT, with front and rear brakes. Notice that the shocks have three mounting positions at the top and three at the bottom, so their angle and leverage can be varied to suit the track. Brake lever is on the right, gearshift lever is on the left (on this example), while most Camel Pro riders prefer both pedals on the right—one above the other, so the left side can be laid on the ground without bending anything. *Cycle World*

made singles or small bikes for years and the Aermacchi program was long over. But it did want to keep the team in contention and the company didn't mind having something its dealers could race or sponsor, as Harley-Davidsons.

Parallel to this, the Austrian company, Rotax, a division of Canadian Bombardier, was making a good, sound, up-to-the-minute series of engines, for sale to all comers. One was a four-stroke single, a 500 cc with four valves, a single overhead camshaft driven by belt, with five speeds, unit construction and so on. Rotax didn't sell motorcycles, although the parent company had a motorcycle division, Can-Am, but it did sell engines to motorcycle companies like SWM and KTM that were too small to make their own.

Model	500-R
Year	na
Engine	sohc four-stroke single
Bore and stroke	3.50x3.126 in.
Displacement	30.1 cu. in. (494 cc)
Brake horsepower	52
Transmission	5 speeds
Wheelbase	53 in.
Weight	260 lb. (dry)
Wheels	19 in.
Tires	4.00
Brakes (factory)	na

The 500-R's debut, as Ricky Graham gets the bit in his teeth and charges to victory in the last race of the 1983 season. Actually, Graham and Tex Peel, his tuner, had this machine when it was a Wood-Rotax and prepared it for the day they could run it as a Harley. Graham and Peel won the national championship as privateers, but had some factory help and cooperation and preferred to keep in the factory's good graces. Graham's ferocious expression here is accurate: He had to win this race while Randy Goss got no points, if he was to retain the No. 1 plate. Graham won going away and lapped half the field, in as fine a TT ride as anybody ever had. But Goss finished seventh, and was the new champion. *Cycle World*

And there was a racing version, in the hands of Ron Wood, a California sportsman, sponsor, tuner and businessman. Wood was famous for doing immaculate work, and doing things his own way. He was the US importer for Rotax engines and Dell'Orto carburetors.

So Harley-Davidson arranged to buy, through Wood, three complete motorcycles and twenty-three engines, all set to racing specs. H-D listed the engines with a part number and certified that the engine could be bought from any Harley-Davidson store. Thus the engine became a certified Harley-Davidson motorcycle, eligible for AMA racing under Class C rules.

Wood had built a number of winning machines for Alex Jorgensen, notably Nortons and a few Can-Ams, Rotax powered, so he was qualified for the job.

Technically, the 500-R, also known as the 500-HDR in the brochures, was a success. Ricky Graham, who with tuner Tex Peel won the Grand National title in 1982 with some factory help but not a full program, defended the plate in 1983 with more help but not team membership. It came down to the wire and Peel and Graham were worried because they needed a competitive bike for all the events and Harley didn't have a full set. They got a machine from Wood, which would have had to run as a Wood-Rotax or a Can-Am, except that Harley-Davidson did the paperwork in time for Graham to ride his newly painted 500-R in the last race of 1983, and win. (Not the title, though; Randy Goss took it for the official team.)

Wood did a good, straightforward job. He built the frames out of small, steel tubing, with the larger central backbone and twin tubes running around the engine and triangulated rear section, plus an extra loop out to the end of the rear fender. Wood supplied the 1.75 gal. fuel tank and the fiberglass seat/fender, and used one of his favorite tricks; putting the oil into the frame tubes so he wouldn't need a bolt-on tank for the dry-sump engine. The machines were delivered with disc brakes front and back, and with the 35 mm Marzocchi forks that were standard wear at the time.

Then the partnership went on the rocks. First, the Harley-Davidson factory and team needed to be on good terms with all the racing suppliers. So when the model was announced, the brochure made it clear that a prospect couldn't get any of this from the factory, but was expected to deal with *either* Wood or the other dozen builders of frames, tanks, brakes and so forth. The 500-R came complete, a complete engine that is, except there was no carburetor fitted. You could get a Dell'Orto. Or you could get a Mikuni, or anything you wanted.

That was business and Wood probably wouldn't have minded that.

But the Harley team had inner partnerships and they were free to make certain choices of their own. All three disliked Wood's frame and the way the bike handled, so they changed to Knight frames and re-did the engine tuning to deliver power the way each rider wanted it. This was reasonable, and productive: Randy Goss won both short-track nationals and one of the season's four TTs in 1984. Nor was Wood's design bad, as Jorgensen won the 1984 Ascot TT, riding for Wood on a Wood-Rotax—same design the H-D team had rejected.

In terms of facts, the team tuners say they liked the Knight frame's engine positioning, weight distribution and steering head angle better, with no details included. Wood refuses to talk, except to detail Harley-Davidson's shortcomings in business and in sport.

Later revision of the rules allowed construction and homologation of an engine identical to a Harley-Davidson, a Rotax, a KTM and a Can-Am, and permitted variations; the same package but as a 560 cc or 600 cc version. For the 1987 season, the AMA reverted back to a 500 cc limit for the singles.

The engine remains a superior engine, at least equal to the Honda single of the same displacement. There have even been some experiments, like fitting two Rotax heads and barrels to a common crankcase, in a vee, just the way the pioneers did it at the turn of the century, but nothing has been made public.

Overview

Our focus here is on the factory team, as well as the production racers and national championship racing.

There have been some other projects, good ones, taking place outside this framework ever since the first H-D two-stroke began the most recent lightweight chapter.

There were some good enduro bikes with rear suspension built out of the rigid Hummers, for instance. Bruce Ogilvie twice won the 250 cc class in the Baja 1000, using modified versions of the MX-250. Before that, there were some winning Baja bikes in the 100 cc class. There were Bonneville racers built with the smaller engines and they did well, along with some road racing done with the 125 cc Aermacchi that was doubled up later into the RR-250. And there were road racers made with the 350 cc ERS four-stroke, the factory's last try in that direction.

Which brings us to the close of this era, and not with much of a bang. In 1979 it was clear to Harley-Davidson management that the whole thing simply hadn't worked, that they might as well stop building small motorcycles and stick with the machines they knew they could sell.

It's hard to fault the decision. If Indian and the English makers had come to the same conclusion in time, and not tried to fight the other man's fight, so to speak, perhaps they'd be with us still.

Financial success or not, Harley-Davidson built some good lightweights and won some races, which is what it's supposed to be about.

Chapter 4

Modern times (1970-86)

One of the few similarities between racing and real life is that disasters never occur when convenient. So, just as the tire goes flat when you're all dressed up for the dance or the job interview, so did the end of the 750/500 equivalency formula arrive when Harley-Davidson Motor Company had heaps of other problems on the corporate plate.

Mostly, it was money. The motorcycle boom was in full swing. Harley-Davidson hadn't caused the surge in popularity, in fact was sort of surprised by the whole thing, but the company lacked the capital to expand and take advantage of what the Japanese had created. The company went public but that didn't raise enough money so the corporate owners and managers wound up in a struggle to be merged into a larger conglomerate, one they hoped would like motorcycles and keep the company going.

That they got in the person, well, corporation, of American Machine and Foundry, which followed another trend and renamed itself AMF.

AMF had money and its management had plans. The chairman of AMF, by great good luck, was a motorcycle enthusiast, Rodney Gott. AMF was ready to move and shake, except that the managers were new and businesslike. They didn't know much about motorcycles and they weren't family, and they were clumsy at making people do what had to be done. As a result, what with moving operations from one place to another, there were problems with the unions and there was sloppy work and lack of quality control and production that expanded too quickly.

And racing wasn't a priority.

Dick O'Brien, Clyde Denzer and the guys on the team and in the racing shop saw this all too clearly, just as they'd seen the end of the KR approaching long before it actually happened. O'Brien had mentioned this to his superior officers, but they'd been busy and at first said to begin planning; then they said to forget it. So while the racing shop and the engineers had been fiddling with a full 750 cc racing engine, or with a 750 they could put into production or build from production parts, they hadn't been able to go beyond the planning and talking stage.

Then the AMA reorganized, and the new committee and new congress changed the rules. For 1969, the dirt-track Class C rule was for production 750s, with road racing held over for 500 overheads and 750 side-valves. The 1970 national championship would be for 750s all the way around.

There was another factor thrown in at the last minute. As mentioned, during the juggling and swapping over rules there had at one time been an intention to limit national races to 350 cc. Some companies, notably the Japanese, had taken this at face value and had begun designing racing engines. The western political system took them by surprise.

So, perhaps to make the Japanese feel better and stay in American racing, the AMA rearranged the lightweight class, for short track and for road racing. For 1970 and beyond, those two classes would be for 350 cc four-stroke singles and twins, 360 cc two-stroke singles and 250 cc two-stroke twins.

The Japanese, in the form of Yamaha especially, found this to their liking. The demands of the two forms varied, of course, but the 360 two-stroke single was the way to go in short track and the 250 twin ruled lightweight road racing.

Harley-Davidson managed to stay in the game for a few more years with the two-stroke single, enlarged for the team only, and the 250 twin from Aermacchi.

The first XR-750

The racing and engineering departments had more of a challenge with the 750 cc limit but they did what they could with what they had.

This was a completely Harley-Davidson-style project. Remember, back when the flathead KH was replaced with the ohv XL Sportster, the racing KHR was replaced by the XLR, a Sportster-based racing engine. The XLR looked like a Sportster on the outside and used a surprising number of XL parts, but inside it was full race, packed full of KR parts scaled and beefed for the stress of 900 cc.

Next factor, politics. The Class C rules in 1968, when the displacement class was changed, called for a minimum of 200 examples; that is, complete machines, to have been produced and to be ready for sale if not actually sold.

Harley-Davidson had been producing the KR and KRTT in a leisurely sort of way for more than ten years, so the natural assumption was to file a statement of intent with the AMA. H-D would make 200 examples of the new racer. No, said the AMA committee, you must make the bikes and make them now, or else. "Darned if we will," O'Brien recalls was the official response from H-D, with management then telling him to stop work. A short time after that, cooler heads prevailed and O'Brien was told to resume work. (He'd

Model	XR-750
Year	1970-71
Engine	45° V-twin, ohv
Bore and stroke	3.00x3.219 in.
Displacement	45 cu. in. (748 cc)
Brake horsepower	62 (as delivered*)
Transmission	4 speeds
Wheelbase	54 in.
Weight	320 lb. (approx. dry; varied with specs)
Wheels	19 in.
Tires	4.00 f/r
Brakes (factory)	none on customer version

*Later team engines, with dual carbs and reworked cylinder heads, delivered 70-80 bhp

never actually stopped planning, although he didn't tell them that.)

But what all this meant was that the team didn't have much time, nor much choice of raw material, with which to fight the new style campaign.

It did well during the 1969 season. The KRTT went out in glory at Daytona, beating the 500s for the

A little symbolism here, as Cal Rayborn on his iron XR-750 looks down to see what the blankety-blank has broken now. Behind him is Mark Brelsford, also on an iron XR. Len and Brad Andres

last time, and the KR held up its end in the dirt, where the newly legal 650 twins and 750 triples should have triumphed. (Sorry.)

The Harley team was exceptionally strong. Mert Lawwill, Bart Markel, Cal Rayborn, Fred Nix and Mark Brelsford all had factory help and did all the winning for Harley—twelve of the year's twenty-five nationals, with Lawwill winning the national title. Thing was, they did it on momentum.

Back in the engineering department and the racing shop, the decision had been made to, first, make a 750 cc version of the XLR engine and, second, to concentrate on the dirt because that was where most of the races were run, and where they could expect to sell most of the machines they'd have to produce.

Engine

Harley used the basic layout of the KR, K, XLR and XL; that is, a unit construction V-twin, with cylinders directly fore and aft, 45 deg. included angle, air cooling.

The XR engine—the name came naturally—had the same four cavities: crankcase in front, gearbox in back, primary drive and clutch housed on the left, and gear case with camshafts and drive for ignition and oil pump on the right. The four cams lined up like baby ducks, one after the other: exhaust, intake, intake, exhaust. The intake ports were in the center of the vee with the exhausts facing right, in front for the front barrel and in back for the back barrel. The magneto mounted horizontally, in front of the front cylinder and driven by gears off the cam drive, just like the KR. And the oil was dry sump, with a small cavity at the bottom of the crankcase and with a two-stage pump pulling oil from a remote tank and pushing it to the cams and the valve gear.

What may have been the engine's fatal similarity was the use of cast-iron barrels and heads, shared with the XLR and XL Sportster.

There was nothing inherently wrong with cast iron, not in the long run. But iron retains heat while aluminum alloy, as used for Harley's big twins and virtually every other motorcycle engine by that time,

The 1970 iron-barrel XR-750, from the press kit. Notice first what a beautifully proportioned and styled piece of work this machine is; second, how much it's like the KR, right down to the cases, covers and exhaust pipes. Those are Ceriani forks, there are no brakes although brakes were legal for dirt track by then and the frame is obviously descended from the KR. *Harley-Davidson*

When the racers got the brakes

The man who gets most of the credit for putting brakes on flat-track bikes says now he's not sure it was that good an idea.

Dick Mann won the national championship twice, in 1963 and 1971. He was his own man, and rode Matchless, BSA, Yamaha and Honda, while never quite riding *for* any factory. Mann thought for himself and worked for himself and was (and still is) highly respected in the sport. Dick O'Brien says Mann would have won the title more than twice if he'd worked for a factory or, more likely, if Mann had been willing to spend the money instead of scraping by.

The brakes, though, were the result of thought and circumstance.

The thought was that motorcycles got more powerful and faster in the late fifties and early sixties. "There was no way to stop in an emergency. I thought that was dumb," Mann says.

So Mann and another racer, a brilliant if moody (and truculent) rider named Albert Gunter, began asking the AMA to allow brakes on flat track. Brakes had been banned since the days of board tracks and banked speedways, and since the AMA always liked to do what had already been done, it voted no, year after year.

Circumstances changed: Two-strokes had become useful, thanks, in large part, to an East German engineer/racer named Ernst Degner, who escaped from behind the Iron Curtain and signed on with Suzuki (which had great success expanding on the work done by German engineers).

Racing became popular, and two-strokes and four-strokes, 250 cc for short track were used in indoor races in arenas and auditoriums, usually on concrete floors with not much traction.

The four-strokes had engine braking; that is, when the rider rolled the throttle shut, the engine slowed and served to slow the bike. Two-strokes had compression on both sides of the piston, so they didn't have engine braking. Consequently, the two-stroke riders began using compression releases, valves that opened half the engine to the atmosphere and thus let the other side's compression act sort of like a brake.

The compression releases worked better than closing the throttle, so the two camps argued back and forth until it seemed sensible to everybody that they be allowed to use brakes. Then, what with the same guys racing the indoor and short tracks and the larger races, nationals or not, AMA or local, brakes weren't seen as a bad thing. The AMA allowed the use of brakes at Ascot, a rough and different type of track, and finally, in 1969, allowed brakes to be fitted to flat-track machines.

The brakes were optional and were used at first, just as Mann had expected, for slowing or stopping in an emergency. Some guys had brakes, others didn't. They weren't used as a racing trick, though, because the racers who were then licensed Experts had been trained without brakes. Like Carroll Resweber, if asked how they did it, they'd look surprised and say, pitch the bike sideways and it slides to a stop, what's the big deal? Do skis and skates have brakes?

Then the new generation arrived. They'd been kids on little dirt bikes, ridden enduros and perhaps even motocross and didn't know motorcycles don't need brakes. They rode with brakes and used the back brake, which is all you'd put on a flat tracker anyway, as a riding technique.

Mann says brakes ruined the half mile as the best spectacle in racing. Not for any terrible reason, not on purpose, and not without the equally innocent help of the AMA and Goodyear Tire and Rubber Company.

Tires? Yeah. At about this time, by coincidence, the rules changed. The side-valve 750s and ohv 500s had been replaced by ohv 750 twins and triples and even fours.

Until then, Class C rules had always required street tires; not merely street legal but really street, the tires road riders used to ride on roads, good for thousands of miles with the modestly powered motorcycles of the day.

The big twins, triples and fours—the 70 and 80 bhp engines—literally tore those tires to shreds. Mile races were being run on tires that would be junk before the flag. So the AMA did the sensible thing, and Goodyear did the proper thing, designed and built the DT series of tires, DT for dirt track, of course. They had tread, they looked mostly like road tires and they weren't knobbies, which the AMA would have banned and always will.

But the DT was a racing tire. It had a stronger carcass and a softer compound because it needed to transmit power and withstand heat, but it didn't need to last beyond thirty or forty miles (leaving a margin of error in case of a re-start).

The soft tire left its compound in the track. Not on, *in*. As the tires wore down, the rubber, which is more a synthetic than rubber, mixed with the dirt, which was as much chemical as dirt. This rubber/dirt mix packed into the famous blue groove, the hard-as-pavement surface that forms a narrow line around the inside of the corners of the modern dirt track.

Why a thin line around the corners? Mann explains: Begin with imagining the half circle that's at each end of a half-mile track. Imagine five points around the half circle. The first point, 1, is right where the straight begins to be the turn; 2 is between the start of the corner and its center; 3 is the center, the apex of this half circle, 4 is between the apex and the beginning of the straight, and 5 is where the turn becomes the straight.

Now then. Formerly, a dirt track wasn't packed. It was loose and smooth, undisturbed by brakes, suspension and tires that could make bumps, ruts and waves.

Riders came down the straight on the outside, on the fence where point 1 is on our imaginary map. They'd cock the bike sideways to slow down and they'd slide in, aiming to just nick the apex, point 3. Then they'd get on the gas and powerslide out again, throwing a roostertail of dirt high into the air and into the stands, showering and delighting small boys of all ages.

Now, says Mann, obviously you can slow more quickly with brakes than by sliding, but what you can't do is brake in a slide. So now, the riders come into the corner, at point 1, on the inside, upright and under braking. They pitch the bike sideways at 2 and slide out, going wide at 3 and getting back on the gas and nicking the inside at 2, and keeping the slide tight and on down the straight to the half circle at the other end.

The immediate victim, Mann says, was his close friend and professional rival, Mert Lawwill.

"I'd retired by then, but I used to go to the races and watch. Mert was the last man with the old-time style. If he had the best qualifying time and won his heat and got the pole, he'd win the race. But in traffic, he couldn't operate," Mann says.

"Mert had a helluva time in traffic. In 1974 and 1975, he was bruised purple all year!"

Why? Look again at the map, says Mann, and you'll see that if you draw an arc for the old way, and an arc for the new way, the two arcs will cross twice for each half circle. If the old style and the new style go into the turn side by side, they're going to bump. If they don't bump going in, they'll bump coming out.

Worse, the art of passing in the turn has become a lost art. Without brakes, everybody could go into the turn side by side, figuring you'd get yours sideways under a guy and get to 3 before he did.

But then, Mann says, the traction was the same across the track. Now it's all grip on the blue groove and all marbles outside, on what used to be the cushion. Of the current crop of racers, only Scott Parker and Jay Springsteen, graduates of Michigan's sandy tracks, really ride the cushion. The others line up and go into the turns single file. They pass by nip and tuck, so to speak.

"Most of the best races were on the half miles," says Mann ruefully. "If I'd known this was gonna happen, I might have had second thoughts."

---

flings heat into the airstream. And the design wasn't the optimum for racing, with intakes in the center of the vee, paired on a common manifold and being fed by one carb, with the rear exhaust port at the rear of the cylinder, sheltered from the airstream.

But this design was obviously a member of the Harley-Davidson Sportster family, and it was nearly identical on the outside to the production engine guys were buying for daily use. It was cheap to build, and it was quick: O'Brien says the chief designer, Pieter Zylstra, had only four months to do the whole thing.

The inside of the XR-750 engine was a lot different. The flywheels were the same as those for the KR and the XLR, smaller and lighter than the XL flywheels. This is something of a speed or racing technique. Heavy flywheels smooth and even out irregular power pulses, for which the V-twin is famous. And they keep the engine going at low speed between power pulses. However, they also impede acceleration and consume power that might be used better elsewhere. In this case (didn't mean to do that), a smaller pair of flywheels took up less space and helped the V-twin's struggle against variations in crankcase air pressure.

The XR flywheels, like those in the KR and the XLR, had separate mainshafts (the shafts in the center of the flywheel, which project outboard) to the main bearings on each side and then to the engine sprocket and primary drive on the left, the gear case on the right.

The flywheels, as on all Harley twins, were joined by the crankpin. The XR got the massive crankpin used for the XLR. As on that engine, the crankpin was held in tapered holes in each flywheel, with a giant nut on each wheel's outer face.

The XL, KR and XLR flywheels held their mainshafts in place with tapered mounting holes (tapered in the other direction from the crankpin mount) and with securing nuts on the flywheel's inner face.

Now then. The first XR-750, known as the iron engine (and with an unflattering nickname as well, but we'll get to that) was in outline a small version of the XLR. That engine was a proven performer, so one would think (and the team seems to have thought) that all it would take would be a smaller version of the 883 cc XLR, one to fit the AMA's new 750 cc limit. So, because big bore equals big valves and more power,

There was some doubt about how many XR-750s the factory actually made, and when, so H-D released this photo of all 200 XRs, lined up in the warehouse and stacked all the way to the horizon. At least half the 200, and a few road-race versions of the later alloy XR, were scrapped for accounting purposes. Harley-Davidson

and short stroke equals more revs and more power, the XR was a destroked XLR.

Keeping things straight here, recall that the KR's bore and stroke were 2.75x3.8125, same as the WR (and the K). The XLR had the same bore and stroke as the XL, 3.00x3.812; so the XR, with stroke reduced to 3.219 for 748 cc, gave the new 750 a different bore and stroke from the old one.

That's just figures. The problem came when the big, thick crankpin moved toward the center of the flywheel to shorten the stroke; it came so close to the inner ends of the mainshafts that there was no room for securing nuts on the shafts. So the taper angle was changed and a keyway and woodruff key provided, and the shafts were fitted to the flywheels with an extra-pressure interference fit.

Back with XLR practice, the main bearings were ball instead of roller, and the camshafts were spun in sealed ball bearings. The engine sprocket was keyed and tapered instead of splined, and there were four choices of gear, with a twenty, twenty-five, thirty or thirty-four-tooth front sprocket, with triple-row chain sized to suit for the fifty-nine-tooth clutch sprocket. The seven-plate dry clutch was retained, although one year later the Sportster clutch was changed to run in the transmission's lube, and to heck with trying to keep a dry clutch happy in a wet primary case.

Racers, of course, maintain their equipment, and because the dry clutch gives more grip for less spring pressure the dry system with seal and cover was retained. (Later, the invention of silicon sealer would make this as good an idea as it should have been all along.)

Camshaft sets carried over from the KR and XLR and sporting Sportsters. Likewise the magneto and the quarter-speed oil pump. Not only did the XR barrels *look like* Sportster parts, they *were* Sportster parts, with half the stroke's decrease, 0.30 in., trimmed off. That's one cooling fin, by chance.

The racing shop was already stretching the spirit of Class C with XLR heads. It would pick the best rough units, and then shift the boxes around before the casting was actually done, so there would be extra material to port and reshape and enlarge. The result *looked* exactly like castings off the dealer's shelf. So the same was done for the XR; that is, the basic heads looked the same, but the XR had larger intake and exhaust valves, and larger ports finished by hand.

The iron XR in action, Rex Beauchamp up. The first XR skated the front end, while the engine wasn't quite as powerful as the English twins were; when it was, it broke down. Harley-Davidson

The pistons for the first engines began life as blanks for the Offenhauser midget engine, a 1500 cc inline four, so using two for a 750 cc twin was sensible because the bores of the two engines were the same. These were classic, high-domed pistons for the XR's classic, high-dome hemispherical combustion chamber.

The XR engine got new cases that looked just like Sportster cases but were different inside because of the different bearings and so on. The gearbox worked the same, carried over from the KR except that the kick start, formerly required, was now an option. The XR cases were stamped 1C1 and dated HO, the eighth letter for the eighth decade, and the 0 for the first year of that decade. Why they changed from the year and the model to this code, nobody remembers.

There *should* be only one engine for every number. You never know, though, because the rules said you couldn't replace the engine after qualifying, which means a spare with the same number could have come in handy, even after the ownership rule had faded away.

This is probably the oddest, rarest iron XR ever built. Note the pushrod covers and rocker boxes barely visible inboard of rider Larry Darr's right knee. This in an ohv Harley engine. But, the air cleaner is jutting out there on the left, like the old side-valve KR, what gives? Lots of machine work. A West Virginia tuner and dealer, known only as Smitty, thought the air cleaner on the right interferred with the rider's knee and balance on the bike. So he swapped front and rear heads, then turned each a half turn so the exhaust ports were at the outside of the vee and the intake ports on the inside, both on the left! Then he modified the rocker boxes because the cams and pushrods are obviously still on the right, and there you are. Didn't seem to hurt, anyway, and illustrates how determined race tuners are, and how the H-D engine can be adapted to circumstances. **Harley-Davidson**

Frame

The rest of the machine, of which Harley had to make as a complete unit, in 200 examples, began with the frame that became the Highboy, the 1967 road-race frame. Not the later one that wrapped around the engine. The new engine was taller than the old flathead, and the extreme bend in the backbone put the steering head too low for use with dirt suspension.

The 1970 frame got a new rear section, with two smaller tubes that branched out and back from their junction with the bigger backbone tube. There were similar tubes angling down to the old-pal twin tomahawks and back to the shock mounts.

Forks were Ceriani, the best of their day and a bonus from Harley's Aermacchi connection, and the shocks were Girling, also tops in their field. The Wixoms, who did the super work on the fairings for the glorious days of 1968, did equally good work for the fiberglass 2.25 gal. fuel tank and the tapering seat base/rear fender with snap-on pad. The 3 qt. oil tank, shaped to conform with the rear triangle of the frame, snugged under the seat. The team color scheme of 1968 was a hit, so the new XR was blazing orange with black and white trim; it looked just perfect.

Harley-Davidson was one big family, so the racing shop built the first frames. Then a St. Louis dealer, Earl Widman, found a contractor to make the rest of the 200 required by the rules (and the competition).

This was a closely inspected process. There had been lots of tricks and deceit over the years, by all parties, so the newly influential factories in the AMA wanted to be sure it didn't happen this time. Extra men were hired to do the assembling, and all 200 XR-750s finally were lined up in the same warehouse for the AMA. Earl Flanders, a respected and trusted referee, came in to count the results, all at once.

Crying towels, all 'round

The XR-750 was first shown to the public at the Houston show, held in conjunction with the races that traditionally open the season, in February, 1970. The magazines also got their first looks then and they were mightily impressed.

There were some hints of what was to come, but nobody spotted them. When Cook Neilson—then editor of *Cycle*, a tuner, a Bonneville and drag racer with, of all things, an XLR—asked how much power the XR produced, he was told "about 62 bhp." Neilson knew the outmoded KR had been peaked and prodded to nearly 60, and this engine had to have lots more, didn't it? He reported what he'd been told, but commented that cagy old OB (O'Brien) never likes to give away any secrets.

If only that had been the case. Elsewhere, the twins and triples from the other camps were growing into 750s and there were rumors that Honda's incredible four, a 750 and legal in Class C, would be raced. Lawwill figured he'd need 70 bhp to keep up with the rivals.

The rules were fairly loose by then, and the Harley team built road-race versions of the XR, with semi-low frames and the good old Wixom-like fairings. By Daytona time the single carb in the vee had doubled, with two Tillotsons attached to reworked and angled intake ports.

But it was to no avail; these bikes weren't radical enough. They used the proven frame, with the engine

Here's the second version of the modified iron XR heads. The stock head on the right has its intake port on the right, but the other head has the stock port plugged and filled, and a new port is installed on the left, so the head can go on the rear barrel and point forward to the left, a cross-flow pattern. Ted Pratt

Here's more adaptability: The top cylinder head is a stock iron XR-750 head. The bottom head has had the intake port, on the right, trimmed and rebuilt into a right angle turn, so it can have its own carb, one of a pair, as with the old KR engine. Ted Pratt

moved a fraction of an inch but no more. They had the good fairing, the four-shoe Fontana brakes and so on.

And they were dead slow. Fastest qualifier for Daytona was Gene Romero on the Triumph triple, 157.342 mph. There were no Harleys in the top ten in 1970, while in 1969 Reiman had been tops at 149, and all seven factory KRs were in the top ten.

Dick Mann, dropped from the BSA team in 1969, won the 1970 Daytona 200 on a new Honda 750, one that never raced again (Honda does that). Mann got a standing ovation, while the best of the team's XRs broke and/or melted during the race. The best Harley in the field was Walt Fulton's trusty KRTT, in sixth.

The Daytona race exemplified the 1970 season; that is, disaster. The official reason for the team's Daytona retirement was piston failure, but the actual causes went far deeper.

First, the mainshafts that had been pressed into the flywheels came loose. This led to mechanical disruptions better imagined than experienced. This is an odd failure, in that the first reports about the XR engine said the shafts were welded to the wheels after installation, indicating awareness of the potential problem. But the shafts weren't welded in customer engines, as I learned firsthand. And the replacement flywheels from the storehouse of Len Andres (Cal Rayborn's tuner) were unwelded. If Rayborn's engines were left like that, obviously so were some others.

Second, too much heat. The iron XR engine was known, not too kindly, as the Waffle Iron: It created and retained heat far beyond what anybody expected. All one could say later was that the XLR engine was run in TTs, at the drags and at Bonneville, so it ran with alcohol fuel or for short bursts of power only. There'd never been a reason to learn about the heat retention under long, hard open-throttle runs.

For the fourth and last version of the iron XR heads, used at Daytona in 1971, the engine got two front heads, with stubs so the intakes were at the right rear and each carb was at a fixed, tuned distance from the valve. The exhaust port on the rear barrel was moved from right to left, to give room for the front intake, and the rear exhaust pipe routed down the left side of the bike. Ted Pratt

The complete set of iron XR heads. From the bottom, the production heads, then the dual carb heads, then the cross-flow and finally the paired fronts. The paired fronts' drains were plugged, and used outside lines to drain oil from the rocker boxes. All this work got the iron engine to deliver, briefly, as much power as the alloy engine gave when it was new. But, all the grief stayed with the factory. None of the modified heads were sold or offered to the public, so only the team and the favored few labored so hard for so little. Ted Pratt

Third, not enough power. Prior to the 1970 Daytona race, Lawwill had confidently predicted 70 bhp from the new engine. By race time, with the first version of the dual carb XR, he may have had 65 bhp or even the 62 that seemed like such a clever gambit late in 1969. After Daytona, it turned out Romero's Triumph triple had 80 bhp, so even the Harley team's modest target, while not met, wouldn't have done the job.

A fourth factor was the union and related personnel problems, in O'Brien's opinion: "We had a clique going in the racing department. We had problems, we weren't getting the job done and I suggested we get the stuff out of there."

Paul Goldsmith, former motorcycle racer turned stock car star, was a partner in Nichels Engineering, in Indiana. Nichels was given a contract to prepare the road-race machines, and later do engine work.

"We moved the stuff to Nichels and they hired Ron Alexander and Walt Faulk and we had one guy working there almost full time. I was down there once a week, every week. Eventually we got things at the plant straightened around and we brought it back into the plant," O'Brien said.

This wasn't a complete handicap because Rayborn had Andres, and Lawwill had himself and a tuner named Jim Belland.

Belland was a salesman for Dud Perkins, the San Francisco dealer, and was also trained as a biologist. His mechanical work began as a hobby. While he had access to parts, he didn't work for the team and wasn't a mechanic at the agency, much like Berndt and Resweber. And, yes, Lawwill and Belland didn't mind doing things better and quicker than the factory team did them.

But that came later. Meanwhile, in 1970 the new XR broke down, again and again. The photo of Rayborn's last KR shows the machine wearing a seat/fender from an XR, meaning the old flatheads were dragged out of their retirement corners week after week, updated where practical. They weren't fast enough, but neither was the new model. And the short-tracker had just been outmoded, too.

In 1970 there were twenty-five nationals; Harleys won seven. And even worse from a seasonal standpoint, when they didn't win, they broke. Defending national champion Lawwill was sixth for the year, new team member Mark Brelsford was seventh. It was

Do you remember Evel Knievel, or that he used to have Harley-Davidson as a sponsor, *or* that he used an XR-750 in the early seventies? Well, here they are, complete with Sportster front brake. (If you ever go to swap meets looking for XR parts, the big, hairy guys behind the piles of stuff will say, as one, "Oh, like Evel Knievel rides?" They will not have heard of Jay Springsteen or Chris Carr.) Harley-Davidson

the worst team finish since the series was created, the first time since 1954 that a Harley-Davidson rider hadn't finished first or second.

Fixes

Time to do what the tough do, as they say. The first step was logical; weld up the shafts in the flywheels. No problem with that.

The next step concerned the private owner. (There weren't many, as I'll detail later.) When the XR engine was announced and submitted for production, it had a compression ratio of 9.5:1 and spark advance was listed as 50 deg. BTDC. The owners manual for the 1970 XR advised the owner to back the spark back to 35-40 deg. and to machine off the piston top approximately 0.25 in., until compression ratio was reduced to 6.7:1. The book included diagrams for the addition of two oil coolers, one between the return side of the pump and the tank, the other between the feed lines and the cylinder heads. It also explained how to make detailed changes and improvements to the oil system so heat would be carried away after transfer from iron to oil.

All this was to reduce heat by reducing power. And it worked, in that the engine ran cooler with the modifications and lived longer, but less power wasn't much of a solution to the problem.

The team and the engineers were hard at work on that. They knew by then that the valves first fitted were too big, so they unmodified the ports and went to smaller intake and exhaust valves, from 49 and 45 mm diameter to 42 and 35 mm, respectively.

A large degree of the piston problem was that the rings, supposedly a low-drag design, cocked, scraped and destroyed themselves; and the friction created the destructive heat. (The first suspicions had been that the heat destroyed the piston.)

In addition, the combustion chamber was all wrong. The high domes used for piston crown and cylinder head meant the chamber was more like a slice of orange peel than a hemisphere. Because the designers were stuck with those XL castings, they had to reshape the combustion chamber by lowering the piston crown.

Zylstra says Harley knew early on that the chamber was wrong so it built a plexiglass model to study the flow, then went to the smaller valves and moved the spark plug around in the chamber's roof.

Finally, the only way to get a useable compression ratio—one that gave enough power and sensible spark advance, the signs of a working combustion chamber—was to machine out the heads, lop off the barrels and run the piston way up inside the heads.

The flatter chamber worked better, and the thinner piston crown was less subject to heat damage. The engine did look different, to the extent that the in-crowd would nudge each other and comment, "Uhhuh, short-rod engine."

According to O'Brien, the rods were stock and the barrels were short. But then, as Cook Neilson learned with the power claims, sometimes O'Brien tells the truth because it's the truth, and sometimes he tells the truth because nobody will believe it anyway. And

Steering head for dirt-track XR, circa 1970. The crossover frame tube technique came from the Norton Featherbed, and it worked. Clamps and forks here are 35 mm Ceriani, the way to go in the seventies. One of the modifications in the XR owners book was to lower the frame, for easier sliding, by cutting one inch off the bottom of the steering head, below the backbone, and welding it back onto the top.

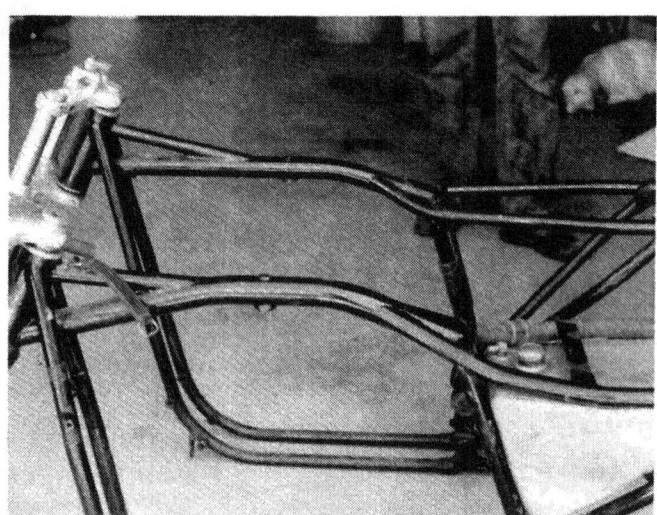

The XR-750 frames were obviously (and sensibly) based on the KR Highboy frame. In the foreground is one of the few, an alloy XR frame with the arced backbone and horizontal brace to the steering head and the 6 qt. oil tank filling the rare triangle. Ted Pratt

because his eyes are twinkling either way, nobody can tell one from the other.

The next step or series of steps to make Harleys more competitive, involved the heads. First came the dual carbs, which did increase power but the tight turn in the intake tract limited the gain.

So sets of heads were reworked to a degree that beggars description. The rear head's intake was moved from right to left, so the inlet for the front barrel faced backward and to the right, and the inlet for the rear barrel faced forward and to the left, a stylish sort of parallel X. The conversion was Draconian, with Bill Werner and another racing shop employee spending hour after hour with grinder, welding and brazing rod and packs of asbestos to keep the heat in the right place. That wasn't the best they could do, either.

The final design was two front heads: The first front head was on the front barrel, with the exhaust port at the front right. The second front head (on the rear barrel) had the exhaust at front left. Both intake ports were at the right rear, pointing backward nearly parallel with the machine, with a long intake pipe and a carb. The whole thing was enclosed in a big air box, with a scoop to feed cool air from the front. It took two skilled men six months to make twelve sets of heads.

By early 1971, with the new season coming quickly, the team also had new flywheels, forged steel and with the mainshafts forged and made as part of the wheel, and with the grain of the metal aligned properly, which you can do with a forging but not a casting.

The iron barrels and heads were still there, but the heads were held in place with long studs through the barrels to the cases, instead of separate bolts from case to barrel and from barrel to head. The engines bristled with oil lines and coolers, three or four per machine.

The factory went to Daytona *en mass:* Rayborn, Lawwill, Markel, Brelsford, Rex Beauchamp and Dave Sehl.

The AMA had changed some rules, and starting position now was determined by lap time, instead of the old, straight run for the clocks on the straight.

H-D wasn't the only flag flying. BSA-Triumph was also out to do whatever it took, with homegrown stars like Nixon and Mann, imports like Mike Hailwood and Paul Smart.

Top gun was Smart, on a Triumph triple. Second was Cal Rayborn.

But. It's *how* they did it. Smart's best lap was 2:09.64, and his top speed through the traps was 150 mph. Cal's best lap was 2:09.79, his top speed 142.85 mph.

How can one man be down 8 mph for the length of a long straight and be 0.15 sec. slower around a two-mile track? Here's how *Cycle* described it:

"Cal Rayborn works the infield like no man alive, his weight on the footpegs, crawling all over the bike, power sliding, man-handling, turning the wheel in

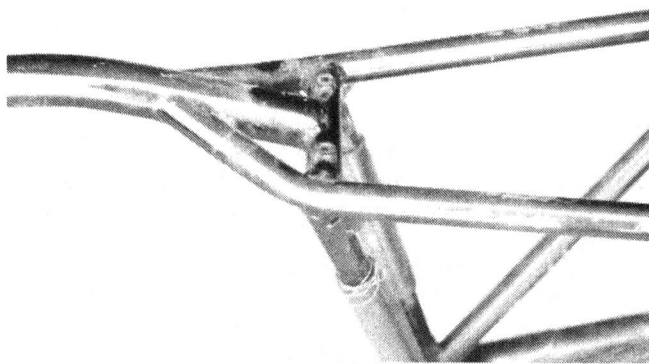

The U of the Highboy and Lowboy frames was changed into this V (or T?) with the backbone, the rear top tubes and the rear engine bay tubes and a cross-brace all coming together as stiffly as possible. The holes in the cross-brace were threaded and took bolts for the seat/fender.

bending and shaping each corner until it complies with his interpretation of it. No one else can push as hard as Calvin."

Cycle went on to say, "Brelsford looks like a younger Calvin . . . maybe, later, a better Calvin."

Hope springs eternal, eh?

But then the bikes broke. Rayborn's gearbox went bad, for no particular reason. When he knew he couldn't win, he wound the engine for the sheer angry joy of it, lapping at 2:08 and 2:09, turning 8700 rpm to

Rear of the flat-track XR frame was nicely triangulated. The swing arm rode in Timken bearings, mounted in the twin tomahawk casting used as a frame member and junction since the KR was designed. The little tabs in the upper left junction were for springs that held the oil tank, and the lugs aft of the shock mount at top right were for the seat/fender mount. This frame, in general, was the design still used and sold by the factory in 1986.

see what would happen. Brelsford's rear sprocket shattered, Smart and Hailwood blew up their engines and Dick Mann, on a BSA that wasn't directly from the factory and thus didn't benefit from the overpreparation that put the team bikes out, strolled home the Daytona winner for the second year straight. Bad luck for some, rough justice for others.

Cycle's final comment was, "After a season of abject misery, the Harley-Davidsons are once again competitive."

Well, maybe. In the next national road race, at Road Atlanta, Rayborn led but his and Brelsford's gearboxes broke before the finish.

And then success; at Louden, New Hampshire, a brand-new track and not at all like the old one except it was in the same neighborhood. Brelsford and Rayborn looked good until Rayborn's oil tank lost its cap and then the oil until the engine dried up and stuck. Brelsford used different lines through the corners, and with a desperate last-lap maneuver, took the lead and kept it.

It was the only road-race national the iron XR would ever win.

At the last race of the year, Rayborn's engine lost a piston in the heat race, then collapsed another in the main event which was run in two parts because of the fear of tire wear. Brelsford's engine broke its primary drive chain. Renzo Pasolini, the Italian star for Aermacchi who came to the United States in a

By the middle of 1970 the team knew the new engine wasn't what they'd hoped for, and they were busy experimenting. This is defending national champion Mert Lawwill with the iron XR in dirt trim. The rear cylinder's intake port has been moved from right to left and gets its own carb, feeding straight back. Mert Lawwill

Harley-Aermacchi swap program, was fourth, best Harley in the field.

That was the last official, certified appearance for the iron-barrel, original version of the XR-750. (It wasn't the last actual race, but we'll get to that.)

Analysis

What went wrong? Several things. We can safely lay blame from here, being so far away from the heat of the battle. But obviously, Harley-Davidson management should have seen this coming, and should have commissioned the 750 project the day after the Competition Congress voted to change the rules.

The engineering department and race team should have set their sights higher, should have known that the Brits had a chance to move fast and build the powerful engines they in fact did build. That leads to a third point, that the engine's design needn't have stuck so firmly on the original castings. If Harley had to make the 200 engines anyway, it wouldn't have been that much more trouble to redo the design by changing the cores and castings and beginning with the heads having two carbs on the right and exhausts on the left they wound up making the hard way, anyway.

Oh, yeah. At the end of the 1971 season, *Cycle* gave a tongue-in-cheek Fancy Footwork award to the AMA official who looked the press right in the eye and explained how those heads met the AMA's definition of stock casting.

But that's hindsight. In the middle ground, the iron engine did win some flat-track races; its later failures were more the result of haste, confusion and luck than of poor design. The primary chain and sprockets and gears are the same in 1986 with 100 bhp as they were in 1971 with 75-80 bhp, only they don't break now.

The guys assigned to make the iron engine work got pretty good at it. Skipping ahead here, while the team was racing the iron engine on the road, the rest of the team back home was designing and building another engine, which we'll meet soon.

At the end of 1971 the full team went testing, with Brelsford and Rayborn, the new engine and the old engine, flat out, side by side.

Ron Alexander and Walt Faulk would like the record to show that on that occasion the iron XR engine, in which they invested blood, sweat and tears, was faster than the alloy XR.

Dick O'Brien agreed, but practical as ever, he said that the iron engine with 80 bhp by then, was at the end of its career, while the alloy engine, brand-new and not even public then, with maybe 75 bhp, was just beginning.

The TT

There were also some highlights from this era, beginning with Mert Lawwill. Most of the top tuners, from Len Andres through Bill Werner, began as not-so-great riders. At least one great rider, Resweber, went on to be a good tuner. But the ability to win races means only that and nothing more. Leonard moved on to win with cars, Reiman and Brad Andres took over the family dealerships, Resweber and Markel once again worked as skilled craftsmen, the trade they'd had before they went racing.

Lawwill (and friendly rival Dick Mann) worked and learned himself into tuning, designing and managing. He drove O'Brien nuts some of the time, but Lawwill's mechanical bent paid off.

It began when Lawwill and Belland built the first of a series of legendary TT bikes. They began with an XLR engine, and did a special frame, which they got away with just barely. They drilled, milled, polished and trimmed the XLR down to 309 lb., forty or fifty less than the other racers weighed.

The rules in the late sixties allowed 900 cc for TT, but most of the racers preferred to stick with their 500 ohv twins, or maybe a 650 on a fast track. The Harley crew generally used the KRTT, or the flathead KHR because it had, well, less power and was less

Bart Markel with the iron-barrel XRTT, Daytona 1970. The new XR used the forks, brakes and fairing from the last version of the KRTT. Daytona Speedway

explosive. So Lawwill won lots of TTs, on the lightest bike with the most power.

When O'Brien signed Mark Brelsford, it was sort of a subcontract. O'Brien was a born coach, tough but fair as the cliche goes, and Brelsford used to worry because O'Brien always seemed to be on his case... until Brelsford realized that almost all of the time, the coach was right. And he cared about the kids on the team.

So, O'Brien made sure Brelsford hung around with Lawwill and Belland. Eventually, they built another TT bike with all the best parts, knowledge and the money.

They, and the fans, called the thing *Goliath*. It was the most powerful XLR that ever turned a corner (Bonneville and drag bikes had more power, for less time) and it weighed less than the 650s and 750s.

Brelsford's fighting weight was about 130 lb. In O'Brien's words, "Mark was the only one who could ride those long TT races at Ascot on that XLR. He didn't manhandle it, he rode it. He had arms about that big [he gestures with his hand, indicating arms

The cross-flow dual carb setup put the carb for the rear barrel in front, and let the carb for the front barrel feed straight in. In exchange for doing what reporters called "great violence" to the spirit of Class C rules, the system didn't work very well. Mert Lawwill

like pipestems], but he had stamina, he could last and last. He'd go out there and ride one of those 100-lappers on that thing, the other people couldn't go 25 or even ten, they'd come in saying 'I've had it.' Mark would go out there and I'll tell you he did a fabulous job, the most fabulous job with that motorcycle anybody could do."

Another event or legend, still spoken of by the people who saw it, took place in 1972, outside the iron engine's official life span.

Two English promoters, Bruce Cox and Gavin Trippe, worked in the United States, putting on races and publishing a newspaper, *Motorcycle Weekly*. They also created a series, match races in England, between an American team and an English team. (The Brits are much bigger on team contests with individual sports, such as speedway leagues, than Americans are.)

The promoters invited Rayborn to be on the American team in 1972, with riders from international makes. Harley-Davidson wasn't officially interested. (It was actually 1971 when the invite arrived; the iron engine had proven to be not good enough, while the alloy engine wasn't official yet, so the team declined.)

Rayborn had his own status, riding for the factory and as an independent, with Andres, so he had a legitimate out. Walt Faulk had his own iron XR-750 engine, paid for and tuned on his own time.

So Rayborn and Faulk and the iron XR went to the match races.

They were a sensation, created a legend not forgotten yet.

But there was something else. On the record, Rayborn won three of the nine races in the series. He set two lap records, but tied with Triumph (and English) rider Ray Pickrell on points because Pickrell also won three races. Therefore, the English team beat the American team on points, so why all the fuss?

Surprise, for one thing. The English knew only that the Americans were cowboys, dirty racers, on crude equipment. And they were fair-weather fliers. So here was Rayborn, not just an American but a Californian! A laid-back, easy-going, winning-smile Californian—a beach boy you could say. But he had that incredible style, a presence the fans could sense from all around the track. And the Harley sounded fantastic; if Faulk couldn't quite keep it on a competitive edge for nine races away from home, well, the bike ran well enough and fast enough to make Rayborn the star of the show.

Scrap iron

There were 200 1970 XR-750s assembled and lined up for official inspection and certification. Even

Cal Rayborn (14) and Roger Reiman, under their bubbles at Daytona 1971. These are the last-try iron engines, with two carbs on the right rear and with the exhaust port for the rear cylinder, topped by a head designed for the front cylinder, moved from right to left, and with the pipe and reverse-cone megaphone on the left side of the engine. Also visible are the four-shoe Fontana front brakes and the rear disc and torque arm to the rear of the frame. These bikes never got a look at the leaders. Daytona Speedway

so, there have been rumors ever since that the factory didn't make all 200 required by the rules. The rumors stayed alive because nobody ever saw all those machines again. The bike didn't have an auspicious debut, after all, and it won only a few races during its first year, so not many private racers bought one. Because they were for racing only and dirt track at that, there wasn't anything else to do with them, so . . . if they weren't sold, and they weren't in the warehouse, where were they?

Scrapped. O'Brien says that during the early seventies when the factory was being organized and money was tight, the accountants came to visit. "They were looking for things to write off, they asked what we had, I said 'we're never gonna sell all this stuff; you need a write-off, help yourself.'

"A lot of the XR parts were standard parts, and there were some that could be returned, but most of it was scrapped and written off."

"How many iron XRs were actually sold?" I asked.

"Around 100, maybe a little more," O'Brien replied.

The 1971 season wasn't even as good as 1970, highlights or no. Brelsford and *Goliath* won the Ascot TT, and Rayborn's iron XR held together long enough to take the mile at Livonia, Michigan, the only dirt-track national Rayborn ever won.

Lawwill lost most of the year to injuries. A tire blew at Daytona. "I remember the first time I hit the ground, then it was strangely quiet and I knew I was in serious trouble because I was in the air again."

That resulted in a broken wrist and elbow. Then at Castle Rock, the TT, his wrist was so badly broken the first doctors wanted to fuse it. Rayborn helped him escape them, then actor Steve McQueen got Lawwill to a sports medicine specialist and it was fixed right.

But Harley-Davidsons won six of the year's twenty-one nationals and Brelsford, seventh for the year, was the only H-D rider in the top ten.

Alloy XR

O'Brien sums it up: "Toward the end, the iron engine was up in the 80 hp bracket and we had a couple of engines that were damned swift. But it was a quickly thrown together thing, without testing and development work, and it surely fell short in some areas. I have to admit that.

"The good part of the cast-iron engine is that we worked so hard together. We learned about cooling, we learned about some of the alloys, we learned about combustion chambers, we learned about every damned thing, and it all went into the aluminum engine.

"It was a poor setup, very much of a crash program and part of the problems were internal problems too, and it was expensive.

"But I think overall, it was a good program."

And in that light, disasters aside, it probably was.

O'Brien knew all the time, though, that there would have to be a real racing engine, both in the

A busman's holiday for Cal Rayborn, who went to Bonneville and climbed into this streamliner, the work of Manning, Riley and Riviera. He'd had a poor season and didn't win any money, he said later, so he signed on for the project. The liner tipped over and rolled over, 35 times on one run somebody said, although they didn't explain how they counted. But when they got that part sorted out, the thing went 265.492 mph, pushed by an 89 cu. in. Sportster engine on nitromethane, and set a world record for motorcycles. Harley-Davidson

spirit of Class C and to the letter. So while the iron engine was being built and raced, and prepared outside the plant, inside they were working on a much newer and different design. That too is part of the tradition.

A logical beginning for the new engine was the redesigned flywheels used to patch up the iron engine. They were forged, with mainshafts and wheels all made as one piece. Working backward here, the iron engine left off the securing nuts on the shafts because the crankpin was so close to the center of the wheels. So, with the forged one-piece wheels, there was no worry about that.

The crankpin could move even closer to the center, the stroke could be shortened and the bore enlarged to keep within the 750 cc limit and allow a higher redline (short stroke) and more room for valves (big bore).

That's exactly how it was done. The new engine got a bore of 3.125 in. and stroke of 2.98 in., oversquare for the first time in Harley's domestic history.

The crankpin was the same diameter, but the old one was tapered and held in place with nuts; the new one was a straight, high-pressure interference fit. An alloy plug was pushed into place on the outside after the pin was in position, as a stiffener and a seal, so oil could be delivered to the rod bearings through the flywheel.

Almost all the principles that applied to the iron XR (and the KR and models before it) were used for the alloy XR: Unit construction, 45 deg. included angle, air cooling, cylinders fore and aft, inline with knife-and-fork connecting rods and so on. The primary drive and gearbox and clutch parts interchanged with the earlier engine. There were ball main bearings and sealed ball bearings for the four one-lobe

Lawwill's 1968 XLR, with frame by Jim Belland, was drilled every place there was metal. It weighed 309 lb. and had 75 or 80 bhp. Rear brake was XL on XR hub, front was Sportster, forks Ceriani and seat and tank were from Lipp Plastics. *Cycle World*

Model	XR-750 (alloy)
Year	1972-80
Engine	45° V-twin, ohv, included angle, air cooled
Bore and stroke	3.125x2.98 in.
Displacement	45 cu. in. (750 cc)
Brake horsepower	90 (average customer engine)
Transmission	4 speeds
Wheelbase	56.75 in.
Weight	295 lb. (dry)
Wheels	19 in.
Tires	4.00 f/r
Brakes (factory)	none; drum front and rear

camshafts, same gear case, dry-sump, quarter-speed oil pump and gear drive for the magneto on the front of the gear case.

But this was a thoroughly modern engine. The cases were alloy, of course, but they were made with more of a stronger, better grade of raw material, more beef to support the bearings, shafts and cylinders. (The larger bore meant larger openings for the cylinders to mount on the cases. The new top *can* be adapted for early XR or XL cases, but it's not worth all the machine work.)

The really new part began with the cylinders, which were made of aluminum with a high silver content, an alloy known around the racing shop as Spaceage Material No. 2. Iron liners for the barrels extended from the cylinders and into the cases, with pegs and sleeves to align everything right.

The heads were made from the same alloy, with the same size valves as fitted to the late iron engines, but with entirely new ports and with an included valve angle of 68 deg. instead of the old engine's 90 deg. This allowed a flatter combustion chamber for the same compression ratio, which means the compression ratio could be higher.

The intake ports were at the right rear of each head and the exhaust ports were at the front left, just the way they'd done it for the iron engine, only too late. This was a good design, with the exhaust ports, or hot spots, in the cooling airstream and the carbs tucked out of the way. The ports were shaped and sized, after hours on the test and flow benches. Carburetors were slide-valve Mikunis, standard wear for racing at the time and ever since, with 36 mm venturis when the engine was first announced. The heads had bronze valve stem guides, and valve seats made of iron alloy shrunken into the heads. Compression ratio at the beginning was 10.5:1, which says something about the new engine's improvement over the old.

The Aermacchi connection showed up first with pistons made in Italy by the firm that supplied the Italian racing team, and with the valve gear design. The new XR had pushrods and rocker arms, of course. The rocker arm ratio—that is, the difference between length on the camshaft side and on the valve stem side—was 1.48:1 and valve clearance was adjusted with eccentric shafts. (They're off-center, so rotating the shaft moves the center up and down and changes the clearance between the rocker arm tip and the valve stem.) This design was almost exactly like the late Aermacchi racing singles, and because it worked there, it was transplanted.

On the other hand, there were a host of detail differences, such as the right main bearing was made one size larger, and the camshafts had to be shorter because the cases were thicker and the gear case cover was carried over. The connecting rods were the same design but a full inch shorter than the earlier XR/XLR rods, partially because the engine was shorter and the stroke less, and partially because O'Brien believes in the theory. The actual camshaft timing (lift, duration, overlap and so on) had to be completely redone because the ports, carbs, exhaust pipes, timing, rods and so forth were all different, while the installation order changed. Remember, the old way had the intakes in the center, while the new engine had exhaust, intake, exhaust, intake. Don't try to swap cams, in other words.

O'Brien, Denzer, Zylstra and Matt Kroll began serious work on the alloy engine about the time the team came trudging back from Daytona, 1970.

Their job was made easier by the success of the cycle parts; the frame, suspension and bodywork used for the iron XR. The engine gave more trouble than power but the rest of the machine, make that machines as in dirt and pavement, did fine.

Early in the alloy XR program the team and the executives decided to offer two models, closely related yet separate. This was an extension of the old KR/KRTT idea, updated.

Ever since the licensing program became Amateur, Junior and Expert, moving up as you earned

Mark Brelsford on the legendary *Goliath,* built by Lawwill and Belland. It was lighter and had more power, and Brelsford made the beast's weight work for him. Here he's about to win the Ascot TT national in 1972. Notice the triple disc brakes. Petersen Publishing Company

points, there had been a good market for flat-track racing machines. But, usually, the AMA pros worked their way up from short track, TT, half miles and so forth, while the road racers began with club events, riding little bikes, then more powerful. But few of them moved into AMA or Camel Pro, and those few didn't do it on Harley-Davidsons.

There was a usable market, then, for the dirt model, and 200 new alloy XR-powered bikes were scheduled for production.

The rules called for certification of the basic package; then the maker or importer (with the maker's help) could get approval for the various bits that were changed with the type of event. By 1972 these changes were major; okay, they amounted to a different machine. So Harley-Davidson cataloged a road racer and, yes, called it an XRTT.

Here are some of the differences, as reported in a 1972 factory brochure.

	XR-750	XRTT
Wheelbase	56.75 in.	54 in.
Seat height	31 in.	28 in.
Dry weight	295 lb.	324 lb.
Steering rake	26 deg.	24 deg.
Trail	3.44 in.	3.63 in.
Fuel capacity	2.5 gal.	6 gal.
Oil capacity	2.75 qt.	4 qt.

Because there wasn't a market for road racers, at least not in the sense of selling to teams near and far, the racing shop actually built between fifteen and twenty-five of the XRTTs.

The frames were different, in that they used the same general design, the double cradle, the cast rear mount and swing arm and so forth. But where the flat tracker had the backbone straight across with the brace arched up, the road racer had the backbone wrapped across the top of the engine, like the late Lowboy frame, with the brace straight to the lower steering head. Like the KR, the road racer could be lower, at the steering head and the seat. The stress from pavement and slick tires is greater, so the smaller, tighter frame is needed. The brochure didn't detail this, but the XRTT came with Ceriani four-shoe front brake at first, then switched to four-shoe Fontana when the Ceriani proved to be grabby. Later, after the factory wasn't really selling the XRTT, Harley and the rest of the world went to disc brakes all around.

For the record, the brochure raises several questions. One dislikes to doubt the facts found in books, but I have owned oil tanks from the iron and the alloy era. The former has its filler on the right, the latter on the left. Other than that, they are identical and both hold three quarts, as measured directly from the can

Just in time for the dirt portion of the 1972 season, the alloy XR-750. Probably the first thing noticed is how similar the alloy version is to the iron XR, and then how massive the aluminum barrels are with their wide, thin and numerous fins. The engineers had learned that lesson well. Harley-Davidson

into the tank. Customer XRs have always come with fiberglass fuel tanks while the factory team and most of the racers went with aluminum years ago. It's hard to compare then versus now, but my own experience with the fiberglass tank says 2.25 gal. is all you can pack in there.

The brochure also says the dirt XR is lighter than the road-race XR, which it should be, although it's tough to get an XR down to 295 lb. Harley must have meant *really* dry—no gear lube or oil, never mind gas. And that might have been a target weight, not quite achieved.

The extra wheelbase of the XR came because it was higher, but the steering head was steeper for the XRTT, so it would turn quicker, and had more trail to keep it straight at speeds the trackers never approached.

But before wheels actually rolled, there were logistical and political problems, as there always seem to be.

The alloy XR was announced late in 1971. The next news came in late February 1972, in the form of an official letter from Harley-Davidson announcing there would be no factory team at Daytona Beach that year. The factory was skipping the event because it couldn't supply 200 examples of the new racer in time.

Below the surface, there was more. Harley had actually produced and collected most of the parts to make 300 examples, but didn't have 200 pairs of cylinder heads, for instance, so the testing program wasn't entirely satisfied. Sources within the AMA let the team and the main office know that this time, in view of Harley-Davidson's good record and good faith, they'd let the bikes be accepted. Only this time, the team stuck to the letter of the law.

They may have done the orange-and-black flag a big favor. In April, two months after Daytona, Rayborn, Alexander and Denzer arrived at the national race at Road Atlanta. They had an iron-engine XRTT. It had plates under the barrels, so they'd had to raise the heads away from the pistons, after machining the heads to get the pistons into the chambers, lowering the compression ratio and reducing heat (and power) again. Rayborn was hopelessly out of the running. Quoting *Cycle*, "To see Cal Rayborn apply his massive talents to a motorcycle that is vastly unworthy of him must drive the H-D brass into fits of periodic apoplexy. If it doesn't, it ought to."

And of course, it must have. Point here is, one, the alloy engine wasn't ready and, two, now we know why the factory brass didn't want Rayborn to go to England. Pity no one realized that Faulk could get more from the engine than they knew it had.

Struggle for success

The happy part of the story begins here: The alloy XR made its first race appearance April 30, 1972, at

The new XR used a frame virtually identical to the iron XR's frame, and it had Ceriani forks but no brake, although rear brakes were common by 1972 and had been legal since 1969. **Harley-Davidson**

the Colorado Springs mile, where Brelsford set fast qualifying time, won his heat and came in second in the main event.

Then Brelsford and *Goliath* won the Ascot TT, followed by the alloy XR's first win, Brelsford up, at the Louisville, Kentucky, half mile. In short order came a half-mile win for Lawwill, back-to-back road races for Rayborn, another half mile for Brelsford and a mile victory for team member Dave Sehl.

Brelsford was the 1972 national champion, and Lawwill, Rayborn and Sehl made the top ten. Not quite the glory years returned, but more like what H-D hoped for.

Just as the iron engine wasn't as bad as the record shows, neither was the alloy engine all that good from the start. The very first engines used special piston rings, which expanded too much when hot, and the exhaust valve guides protruded into the airstream and got hot. Both were easily fixed.

The next problem was more fundamental. The rocker gear came from the Aermacchi single. It worked fine there, but when it was adapted to the V-twin, somebody took a shortcut.

The engine's vee is an angle. The camshafts aren't quite aligned with the crankshaft's center, and the valves form an angle with each other and with the cylinder bore. What we have is geometry—angles where the rocker arms start and stop, and the arc they move through.

The engineers designed two variations of the rockers. There should have been four. Besides that, the tappets broke because the rockers weren't traveling right. So another step was added, in which the ends were cut from each rocker arm and rewelded, each with a slightly different location and angle, and each carefully marked as front intake, rear exhaust or whatever. Several years later, the factory's patterns were modified so the rockers came in sets, with the correct angles built in. At first, making the engine *right* meant remaking or buying outside parts.

Some time after the engine got hours of track time, the team discovered that the ball main bearings were breaking up.

This wasn't surprising. The flywheels held in alignment by the crankpin have always been difficult to align perfectly. And, no matter how tight the fit, there's bound to be some flex, which is hard on the inflexible ball bearings. Earlier, with the flathead WR and KR engines turning lower speeds, this meant replacing the main bearings every few hundred racing miles. But with the added stress of higher compression and revs, the XR engine wasn't lasting even that long.

The cure came from Germany, in the shape (literally) of a new design—a caged bearing with rollers shaped like little beer kegs, not quite ball and not quite roller. They were low-friction, like the ball bearing, but the outer and inner cage could swivel, adapting instantly to any misalignment of the crank-

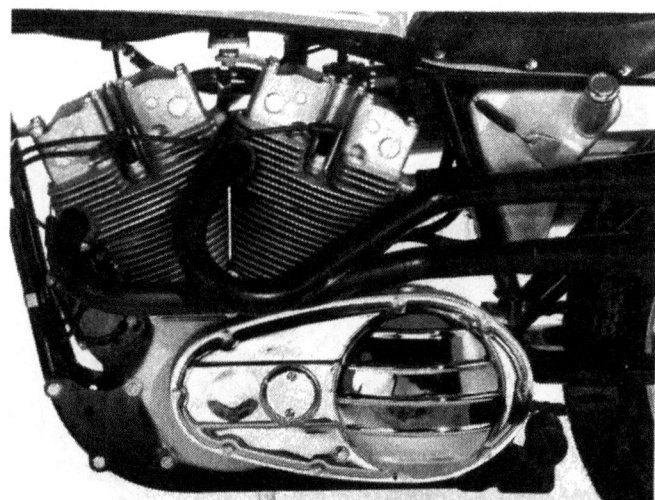

Alloy engine's exhaust ports were at the left front, so the hottest part of the head got the most cooling air. Chrome-plated stamped-steel primary cover interchanged with iron XR and even KR or XL covers. Take note of the front stay between front head and the frame, and the length of the front engine mount plates. Harley-Davidson

pin or the cases. So these bearings, known as Superblend bearings, were a running change. A few years later, for the last production engines, the left-side main bearing was of the same design only wider, known as *doublewides*. (These miracle devices were a retro-fit for early alloy XR engines. With one size smaller on the right, they go all the way back to the 1952 KR. Bless Harley-Davidson's no-obsolescence policy.)

Team tests made public in 1978 and published in *Cycle* show the first dyno runs of the still-secret XR to

Twin 36 mm Mikuni carburetors tucked in and faced backward, out of the rider's way. Gear case cover had a cap over the mounting flange for the magneto, which on this engine has been put on the front of the gear case. Later, the mount and the gear drive beneath it were used for a mechanical tachometer's cable. Harley-Davidson

Two tuned exhaust pipes were cut to a carefully determined length, then given a calculated taper on the megaphone, with reversed cones on the end. This design can give maximum extractive effect, at the expense of a narrowed power band. Harley-Davidson

Alloy engine's hemispherical combustion chamber.

have produced 69 bhp at 6800 rpm. Minor adjustments showed 73 bhp at 7600, with the average team engine delivering 83.9 bhp at 7800 rpm by the end of 1973. There could be as much as 3 bhp, plus or minus, from engine to engine; racing engines have personalities of their own.

The 1973 season was the flip side of the coin. Part of the bad news comes from within, and part comes because the other chaps were doing great work, work so good they'd be punished (not by Harley-Davidson) for doing it.

This was a period of major change. The old rules were so clearly abused that the requirement for approval of frames, brakes, suspension and so forth was abandoned. This was good in that the Honda disc brakes immediately adopted by Harley and everybody else without good brakes of their own, were better brakes. But it was also bad, in that the noble or goodhearted notions of having racing improve the breed took another knock.

The big twins and, more so, the threes and fours had so much power that the AMA dropped the DOT-approved tire rule and asked the tire companies to come up with *real* racing tires. Which they did; the Goodyear DT and DT-2 in particular. Brakes were legal and on their way to becoming a racing technique,

Cutaway model made for display at Harley's racing shop museum, showing the three-row primary chain and the seven-plate dry clutch behind the aluminum cover. Clutch release was by a series of four small pushrods working off a worm gear at the output sprocket. It was the same design used for the 1952 KR, and it worked. Harley-Davidson

Domed pistons are relieved for valve clearance.

and suspension was universal. Racing was different now and a new crop of riders was taking advantage of it.

There were three top newcomers, with the top of the trio being Ken Roberts, followed, make that pursued, by Gary Scott and Mike Kidd, all moving into Expert rank for 1972.

Roberts was a genuine cowboy, a tough and hungry kid from the central California sticks, right off the farm, where he did his share of breaking horses. His early motorcycle rides were Spanish, then he was picked up by Yamaha and he had a good time making fun of Harleys. Roberts was hungry, awkward, rude to the press and possessed absolutely no skill for small talk.

He won his second race as an Expert, the Houston short track 1972. Along with his other virtues, Roberts had no respect for tradition.

Team Harley-Davidson began 1973 with Mark Brelsford as No. 1, teamed with his brother Scott, Lawwill, Markel, Rayborn, Sehl and Beauchamp. For Daytona they had an exchange racer, Italian champion Pasolini.

What they didn't have, was power. The alloy XR was good, but the threes and fours were better, especially the two-strokes. The best Harley was Rayborn's, qualified tenth.

Then Brelsford collided with another Harley, a private entry ridden by Larry Darr, who was having mechanical difficulties. Brelsford's bike burst into flames seconds after he was thrown off. It's hard to say he was lucky, with a broken arm, leg and wrists, but he was. Then Rayborn's engine seized and he broke his collarbone and some ribs. Jarno Saarinen won the Daytona 200 on a Yamaha, followed by team manager Kel Carruthers and Yamaha privateer Jim Evans.

The only fun the team had in 1973 was a trip overseas, to race in Europe and the Anglo-American match races. Harley and Aermacchi reckoned to win some titles (and did) and there were plans to swap more machines.

Brelsford was still out, and Rayborn was still banged up, so he won more fans than races and the

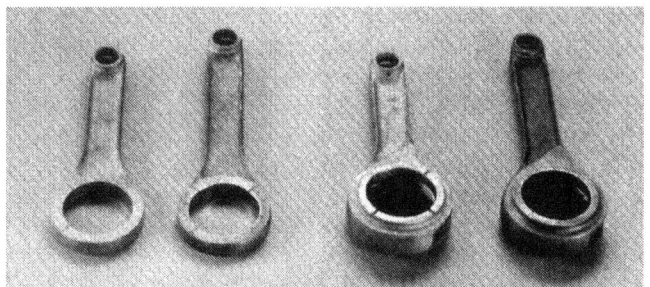

Shorter stroke allowed shorter connecting rods.

Brits topped the points again, but it was all in fun. In Italy, with no one to speak the language, says O'Brien, "we'd go into a restaurant, because I had white hair I guess, they'd take me back in the kitchen, open the oven and I'd say some of this, some of that and they'd say 'Si, Si.' The guys would ask 'What'd you order?' and I'd say 'I dunno.' They'd bring it to the table and everybody would have a good time."

O'Brien continued, "Calvin was fast but he'd separated his collarbone. Pasolini, who was the big draw over there, he was sick and they wanted Calvin to ride his bike with Pasolini's numbers on the leathers. They said that would mean a difference of thousands and thousands of people being there.

"That bike went into a full wobble with him, he had to put so much force into it that it separated the fracture. He rode the entire Easter series with his collarbone broken."

Back home, Sehl and Beauchamp won half miles, Lawwill won a TT and a mile. Roberts had a short-track machine, which Harley didn't, and he had a competitive road racer, which H-D didn't, not quite. Yamaha had begun working out the conversion of its street 650 cc vertical twin into a 750 cc tracker. And Roberts was surely the best all-around rider in the series, so he won the national title for 1973, followed by Gary Scott and Gary Nixon, with Lawwill, steady as ever, fourth for the year.

The last four road races of 1973 went to two-strokes, three to Kawasaki and one to Yamaha. Even Rayborn was out of the hunt, healed or not. *Cycle's* Cook Neilson commented that Harley-Davidson was

Iron liner goes in aluminum barrel.

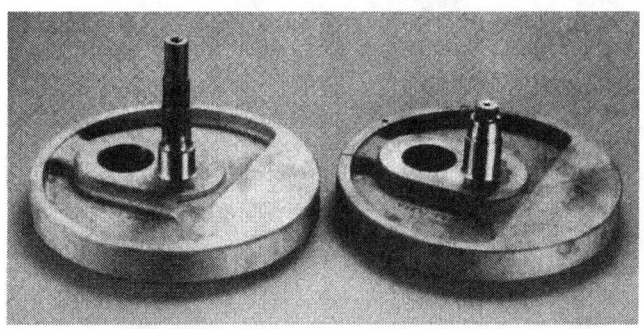

Forged flywheels have integral mainshafts.

"apparently at the end of its rope with the aluminum XR-750 as a viable road race engine, and was possibly pre-occupied with 1974 considerations."

So was Rayborn. He'd never been very happy on dirt, as mentioned. While the wild and crazy dirt racer can control himself in the precision of road racing, witness Roberts, Spencer, Lawson, Mamola, the precise road racers never manage to loosen up enough to conquer the dirt.

To make a sad story short, Rayborn left Harley-Davidson, good wishes on all sides, and went to New Zealand for a series of races during their summer season. The engine of his privately owned Suzuki seized, and Rayborn was thrown off and killed, December 29, 1973. Len Andres parked the old KR in a corner of his shop, next to his (Rayborn's) trophies, and made sure Rayborn's wife and kids didn't go hungry.

Earlier in the season, in Europe, Pasolini and Saarinen were killed in a crash. It wasn't a good year.

The new XR-750 worked well, right from the start. This is Brelsford, on his way to the national title in 1972. Harley-Davidson

The TZ700

Here begins some rules changes with political overtones, changes that had the most impact on American racing since the invention and adoption of Class C. There is injustice here as well, but it wasn't done to, or by, Harley-Davidson.

The protagonist was Yamaha, at home and in the United States. The Yamaha 350 cc two-stroke twins were great racing machines, able to meet and defeat four-strokes twice their size. But when the other Japanese factories came out with 500 and 750 cc twins and triples, the 350 wasn't enough.

Yamaha doubled its engine, literally, into a 700 cc four.

Now then. When the American rules were drawn up, they were prepared by guys from small (relatively) companies. Harley, Indian, Norton, Triumph et al weren't industrial giants; you could say they thought small. When they set the production requirement of 200 machines, up from the previous 100, and before that twenty-five, they did it with faith that nobody could top it. To make 200 machines was to ensure they'd be production bikes, road goers, because nobody could be so eager, or so rich, as to crank out 200 full-bore, true-blue racing motorcycles.

Yamaha was.

Yamaha did.

There wasn't anything personal or nationalistic about this. It was more like sibling rivalry. Yamaha and Honda had conducted their own war, at home and then around the world, since going into the motorcycle business. They'd battled in Grand Prix during the sixties, then dropped out.

Yamaha saw a chance to dominate racing in the United States and in Europe, where Formula 750 was beginning to overshadow the classic 250, 350 and 500 cc GP classes.

So Yamaha made 200 two-stroke fours. The AMA rules said they had to be available to American buyers, so 200 were shipped to the United States, duly counted and certified. The rules didn't say *where* they had to be for sale, so after Earl Flanders counted all the crates, Team Yamaha boss Ken Clark kept four machines and shipped the rest to all parts of the world. That was legal. But, at the same time, more than four orders came in, so some of the bikes came back for sale to the public, just as the rules dictated.

None of the other racing teams or factories had expected or prepared for such a thing. Harley, Suzuki, Kawasaki and Honda were willing and able to make a handful of racing kits and parts based on production machines, but this was beyond anything they could match.

So the AMA changed the rules and made the production requirement one complete motorcycle and at least twenty-four engines, all the same, and offered for sale. No warning, no debate. All the other racing factories were allowed to get even by doing what Yamaha hadn't been allowed to do.

This meant, first, that real production-based bikes didn't stand a chance in the road races and, second, that the road and dirt machines were going to be aimed in different directions from then on.

Meanwhile, for 1974 the Harley team was Mark Brelsford, whom everybody hoped had recovered; his brother Scott, who won a mile late in 1973; Beauchamp; and Gary Scott, co-rookie with Roberts in 1972. Markel had retired to go to work for General Motors, and Lawwill had retired to go into business for himself, making parts and tuning.

Brelsford looked good at the opener, the Houston TT; in control of the uncontrollable *Goliath*. Astonished reporters called it a Sportster and goggled that Brelsford went cowtrailing on the beast. He won a heat, and was eighth in the main, with Scott fourth on an XR. It was the XLR's last national.

At Daytona, all the four-strokes were 20 mph too slow and the XRs were 25 mph too slow. The race was cut from 200 to 180 miles as a sop to the, yes, Energy Crisis. Giacomo Agostini won for Yamaha, the third in a string that would run unbroken until 1985.

Elsewhere about this time the FIM (the European AMA so to speak) declared the Yamaha TZ700 illegal for Formula 750. The promoters promptly revised their rules, making the races for eligible Formula 750 machines, plus TZ700s.

The word that comes to mind for the rest of the 1974 season is . . . slugfest.

The sluggers were Roberts and Scott, each ably seconded by members of their respective teams. Yamaha had the road racer and won five of the six road-race nationals to one for Suzuki. Yamaha had a good short tracker and took one of the two short-track nationals; one for Honda. Harley had neither type of machine and the score for the year's twenty-three nationals was Yamaha 11, Harley 8 (Roberts got six of the eleven all by himself).

Scott won three nationals—one mile and two TTs—all on the XR-750. Harley's other five wins were miles and half miles, where the XR and the newly approved Yamaha 750, built by tuner Shell Thuett from Yamaha parts and with factory help in the casting department, were a close match. In the end, versa-

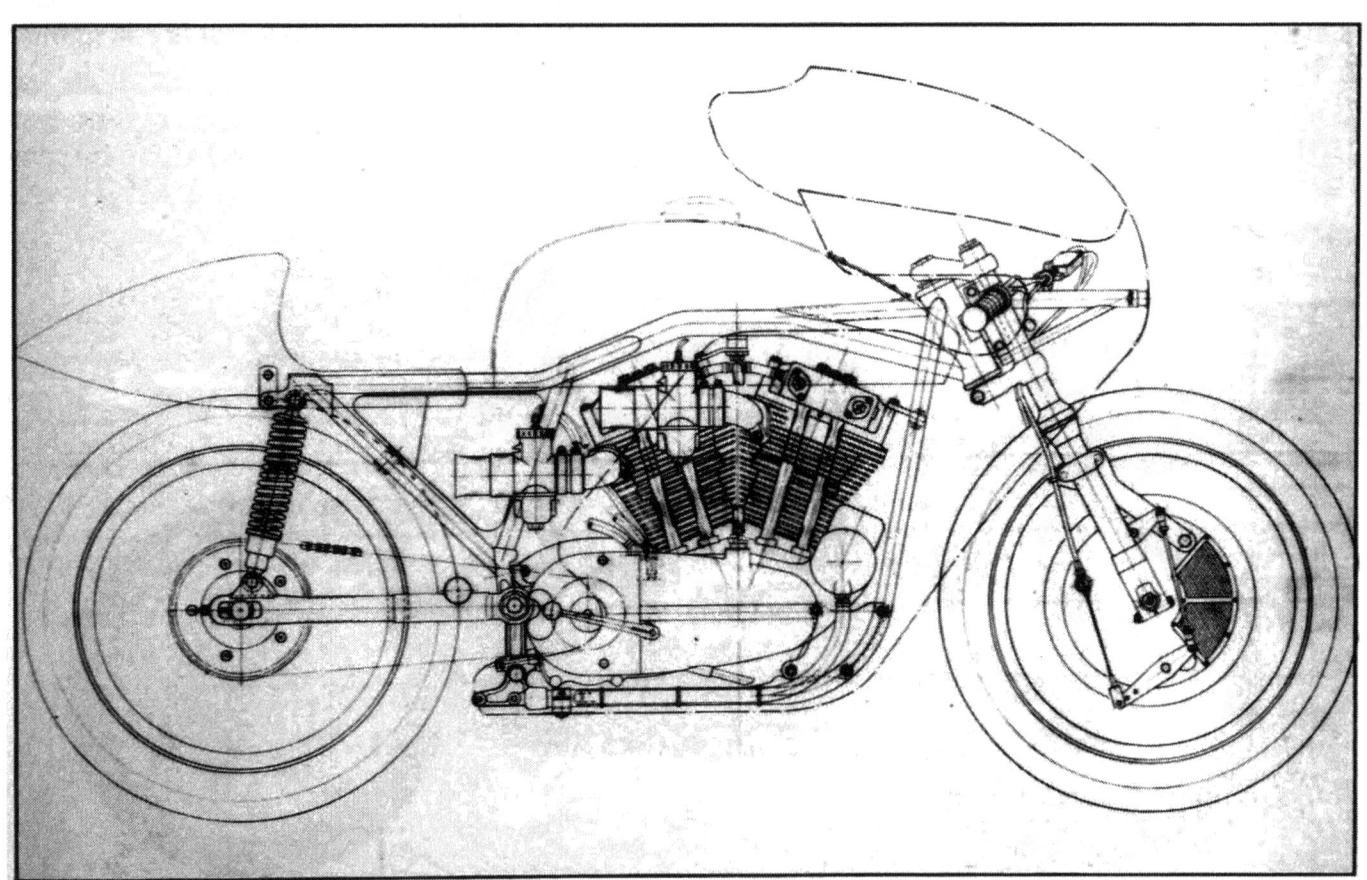

Blueprint of the XRTT of 1972 shows several clear lines of descent. The frame was lower at the seatbase and the rear tubes, and the rear of the backbone was more slanted than the dirt-track version of the frame but the front brace was horizontal and the backbone slanted down, not quite as much as the KR Lowboy, but almost. **The wheels were 18 in., the rear shocks were shorter and the Cerianis had been moved up, so the bike slid down in the front clamps. Harley-Davidson**

tility and Roberts' massive talent won, and he was national champion again.

The XR gets better

Keeping the record straight here, there were 200 dirt-style XR-750s made in 1972. Various people remember the next figures differently, but on the best evidence, the first XR road race frame was made by Ralph Berndt, the tuner behind Reswebber. Then came ten hand-made frames from Nichels Engineering, then forty machine-welded frames from Nichels. Maybe. And this statement has been corrected severely twice already, so it may have to be corrected several more times but, anyway, nobody ever saw more than twenty in one place at one time.

Some of these road-race frames were scrapped when the accountants needed equipment to write off (same time the extra 100 or so iron XRs were scrapped). But the company did find enough buyers for all the dirt-ready alloy XRs, while there were only a handful of buyers for the road-race models. On the other side, there were thirty Yamaha TZ750s entered in the Daytona 180.

So Harley-Davidson was effectively out of the running in the big, fast, long road races. The XR engine just plain didn't have the potential to compete with the two-stroke multis on pavement. (The company and team did try with the Aermacchi-designed two-stroke twins, the RR-250 and the RR-500.)

Instead, the team concentrated on the dirt, and found some useful ways to build better engines.

The first giant leap involved the V-twin's odd problem with crankcase pressure and oil delivery. Even back in KR days, the racing engines had trouble because the pistons pumped air below their tops while making power above, and the crankcases would fill with oil.

The production engine and its racing versions had both used a timed breather, rotating in sequence with the pistons going up and down, and had for years.

O'Brien says, "We wanted to get away from the timed breather because the timing can't be perfect at all rpms.

"To get the old pumps right you had to blueprint everything and scribe the opening and closing points and degree it in with a big wheel. You've got a quarter-speed pump and it's hard to degree it right. Even when we thought we had it, we had two engines at Daytona, they'd both been degreed the same, one would fill with oil and the other was dry as a bone.

XRTT fairing, tank and seat were virtual carry-overs from the KRTT. Air horns have replaced the air cleaners and filters here because there's no dust; otherwise, the pavement and dirt engines were the same. Harley-Davidson

When they were dry, they'd carry one half to one ounce of oil in the sump. Next engine you'd get three or four ounces and it wouldn't run, it would just drag all the power out of it.

"We brought the engines to the plant and tied in manometers [pressure gauges] all over them. The only thing that showed up was, the engine that ran the driest had the lowest vacuum and the lowest pressure."

So Harley stayed with the timed breather and added another breather, topped by a one-way valve, ball and check, out of the gear case. The early XRs ran that way and worked, most of the time.

Then Bill Werner devised a mini-sump, a way to open the bottom of the crankcase, at just the right spot, attach a collection box, with baffles to collect the oil from the flywheels, and route it to the scavenge side of the pump. The idea and initial hardware came from Werner, who spent most of his time, on the job or at home, thinking about related problems, and who (like other top men in this sport) personally owned an XR-750, so he could fiddle with it uncriticized.

O'Brien put the mini-sump idea through all the tests he could devise, then approved it for team use and incorporated it as a running production change for customer use as well.

The next leap was much more obvious. Ron Alexander was working on camshafts and timing. The early alloy engines used cams that proved a bit radical, but Sifton fixed that. Then they used a version of the iron engine's mid-range cams, called the A cams but labeled E in the alloy engine. Next came the BD cams, named for designer Bruce Dennert. Then Alexander worked out the best system with the help of outside designer Sig Erson and some lifts and sequences Erson had developed for the Chevrolet V-8.

While this was going on Alexander discovered that the alloy engine's exhaust ports were wrong. They formed a bottleneck, where the exhaust gases backed up, and in so doing transferred extra heat to the head itself. So the exhaust port was revised, and Alexander went to a larger valve and built an engine for Dave Sehl that had 89 bhp where the previous engines peaked at 82.

These modifications also became team property and were shared with Lawwill. The engines built along these lines, also using longer exhaust pipes and less spark advance (by the way, in case you're tuning an

Footpegs and brake pedal had been moved so far back they were mounted on the swing arm, and the fairing's edge had been trimmed back for clearance. The body- work was a tight fit. Drum brakes were standard issue in 1972, but the team switched to Honda discs as soon as they could get them. Harley-Davidson

When the XRTT was announced, the factory planned to build ten for the team, then 40 more for customers. But rules and circumstances changed; the twins weren't competitive and the production frames were scrapped a few years later. Harley-Davidson

alloy XR in your spare time, don't use 44 deg. the way the book says, use 32 deg.) were called the San Jose engines because they were ready in time for the San Jose Mile. And not a moment too soon.

For the record, there were another 100 XR-750s made in 1975, and sold, presumably for the dealer net price of $2,395. These were based on the 1972 dirt XR, the original frame except the steering head was raised one inch and moved back 5/16 inch, and the top mount for the rear shocks was moved forward one inch.

These figures are much more important than they sound. The fifty frames built for road racing in 1972, the ones made in the shop by hand at least, had lower steering heads and peaked backbones, and had longer swing arms and shorter cradles to put more of the engine's weight on the front wheel. An inch here, a fraction of an inch there makes the difference between road and track racing, and between winning and losing.

The most dramatic Grand National series was in 1969, when the good old side-valve KR beat the newly upgraded ohv 650 twins, and when the 750 cc Triumph triple wasn't raced because it wasn't as fast around the mile track as the lighter, less powerful twins were.

For San Jose, 1974 *Cycle World* took a radar gun and clocked the contenders down the straight. Lawwill's XR was fastest, at 121 mph, followed by Gary

Team Harley-Davidson and guests, at Ontario, 1972: From left, Mark Brelsford (87), Dave Sehl (16), Cal Rayborn (14), Mert Lawwill (7), and Renzo Pasolini (2). Sorry the light's not good, but if you peer closely enough you can see the dual front disc brakes for Brelsford, Rayborn and Pasolini; the four-shoe Fontana for Sehl and Lawwill. The choice of brake was up to the rider. Pasolini finished third overall, Brelsford was fourth, Lawwill sixth and Rayborn crashed. Harley-Davidson

Looking for a thankless task? Try keeping racers in line

Poet Randall Jarrell once wrote that when we look at the pictures of previous generations—those sepia prints and faded snapshots—we can't help thinking those quaint people *must* have known how quaint they were.

There's something of that here, in that motorcycle racing has been big business for nearly a generation now. There are professionals running the show and issuing the press releases, and the riders are trained athletes, dedicated to their sport and perhaps just a bit dull because of it.

Thus, when we look back at the relationships between the guys who rode the bikes and the guys who wrote the rules and tried to enforce them, the view is from one culture to another.

It's sometimes hard not to laugh or not to feel as if they must have known how quaint it all was. But things *were* different then. A bulletin from the AMA to the referees in the field, for instance, ordered that "referees *must* stop listing competition riders' nicknames on their reports."

And they reminded everybody that the AMA rule book had the clause, "Women will not be permitted to compete or participate . . . where speed is a determining factor." (The rulemakers must not have known that back in the 1930s, Dot Robinson, founder of the Motor Maids, rode with husband Earl and by herself to countless enduro wins, and a few cross-country records.)

There did seem to be problems. In September 1955, the AMA notes reported four riders suspended for fifteen days each because their footpegs and levers didn't have rubber covers.

Riders were reminded that each would get one lady's pass and one mechanic's pass for the grandstands, that the mechanics must wear whites in the pits and must change out of whites before going into the stands.

Referees wore striped shirts, and starters wore white. There had been reports that some officials had switched roles without changing shirts.

In October, things were worse. Two riders were suspended for striking officials, and there was a report women had been seen pushing bikes to the starting line.

In 1956, five years after production ceased, the bulletin reminded everyone that it was illegal to use a WR engine with KR barrels on it. Then the AMA published the names of the cheaters, the guys with oversize engines and KR barrels.

Later that year, one rider was suspended because he was caught shifting gears (gasp) during a race. (Dirt machines weren't allowed to have brakes, and downshifting was a way to slow down, so it wasn't allowed for years. There's a suspicion that this began when the imports had foot shift and the domestics slower hand shifts, but that's just gossip.) As clarification, the rule book was quoted: "Once a rider shifts into high gear, he cannot shift gears to reduce speed."

More seriously, though, there was also a list of racers suspended from AMA events because they'd raced in non-AMA events.

In 1957 there was another reminder—no KR barrels on WR cases.

Not mentioned in the AMA bulletins but pertinent to the times, was the feud between the AMA and Floyd Clymer, publisher of *Cycle* magazine and a former racer, dealer and race promoter. Clymer delighted in running polls of his readers, which generally revealed that the readers didn't like the AMA.

On one occasion the poll showed that not one AMA member knew who had elected or appointed the AMA's top man, E. C. Smith, with whom Clymer fussed for something like thirty years. Nor would Smith, a man with faith in himself and his opinions, reveal to Clymer whose employee he was. (Also worth noting here, when Smith retired in 1958, to a standing ovation from the riders at Daytona, Clymer's jabs at the AMA lost much of their punch, somehow.)

Clymer redirected his arrows. After Daytona's 1960 race he wrote, "It is now more than ever a confirmed fact that 500cc overhead valve bikes cannot possibly expect to win at Daytona." And sure enough, a 500 didn't win Daytona until two years later.

The referees meanwhile were reminded to check those ownership papers—this in 1961, when the fiction of the owner/racer was still law.

The next bulletin reminded clubs that women—wait a minute, the actual word used here is *girls*—weren't permitted to compete in scrambles or drag races.

A good move, made during the AMA Technical Committee meeting in 1962, was allowing ads to say "Professional" motorcycle racing. They had been saying Class C, which as the minutes reflect could lead the public to think it was less than first rate.

Oddly though, in 1963 there was a notice that ads weren't allowed to say that scrambles are motorcycle races. Instead, they could be advertised as motorcycle scrambles races, a distinction no longer clear.

Moving ahead of the crowd, for once, the Competition Committee voted in 1963 to allow women—they used that word, not girls—to compete in any AMA event except for Class C.

Speaking of girls, there also must have been some boys in the sport, as the bulletin needed to warn against allowing helmet painting such as Mickey Mouse.

And there it ended. Not the foolishness, exactly, but *Cycle* got new owners and managers, and the AMA went democratic, with elected officials and a different form for the Technical and Competition committees and *Cycle* quit telling the public what the AMA had done, and the AMA hired professionals to issue bulletins.

No more Mickey Mouse, in other words. Seems a shame, somehow.

Pasolini at speed, at Ontario. This was something of an exchange program and Harley and Aermacchi each sent team members back and forth. **Harley-Davidson**

Dick O'Brien, front left, and Mert Lawwill, Ontario 1972. The new XRTT worked well enough but didn't lend itself to the fast, featureless track, nor did the track serve racing very well. It was a dull place to watch a race, and when racing, it was easy to lose track of where you were. Later the track was replaced by houses. **Harley-Davidson**

Nixon and the Kawasaki H2R, a 750 cc two-stroke triple, at 118. Gary Scott and his XR tied with Ken Roberts on the Shell/Yamaha 750 twin, with 117, and Rick Hocking with a Honda 750/4 in a dirt-track frame was clocked at 115.

Roberts won that 1974 mile, and of course the series title, for the second year. Ken Clark, Yamaha's racing manager, says the Yamaha twin, a vertical 750 cc engine based on Yamaha's long-lived 650 twin, had 80 bhp late in 1974, so the Harley and the Yamaha twins were on par. Roberts won because he rode well and because both twins were faster down the straights than the triple or the four: As *Cycle World* intoned after that 1974 race, "Non four-stroke multis have been tried in the dirt before . . . and all have failed."

The 1975 Grand National series was a big surprise, on several fronts. Mark Brelsford had retired: He came back and rode hard after his Daytona crash but never felt right, and speed didn't come as naturally as it had. Then he had another crash and decided it wasn't fun anymore (he'd always done it for the fun), so Brelsford quit and went back north to run a truckline and ranch. (Like many agriculturists, he used the truckline to pay for the ranching.)

Roberts had a complete line of machines, including short track and road race, TT and flat track. Yamaha won the Daytona 200. Harley-Davidson didn't even bother to enter the big race. Instead, Scott was second to Roberts in the lightweight contest.

What it takes

Harley-Davidson had a different kind of stature, or depth. This may sound foolish, but be patient while I explain.

Some people say the 45 deg. V-twin is the perfect dirt-track configuration. The irregular power pulses, their theory holds, give a staggered rhythm and gears the rear tire to the track, with more traction than you can get from any other design, particularly a vertical twin or a boxer or a wider vee.

I know it sounds silly, and there's not a physics professor in the world who wouldn't say, "at 8000 rpm? Geared so the rear wheel is spinning 2000 rpm, you expect this 45 deg. makes the big difference? Get out of my classroom!"

But. By San Jose 1975 the alloy XR engine was nearly in its final form, and it had close to 90 bhp. So did the Yamaha, with special heads made in batches of twenty-five, with optimum ports, valves and so on.

Harley-Davidson depth came in odd ways. For example, there was a strike at the plant late in 1974, and Werner, who was a dedicated union member and a thorn in management's side (when there were sides to be taken), on this occasion took his work home with him. He and Scott worked out of his (Werner's) basement, using their own machines. To test they went out and had drag races. Werner used to race so he could handle that part fine, but because he was a lot heavier than Scott, when they raced Scott got to carry a set of flywheels in his lap. Honest.

Here's another example: There was a t-shirt company in Flint, Michigan, named Vista-Sheen. The owner, Rich Gawthrop, liked motorcycle racing and backed a tuner named Jack Klingsmith, who had ideas of his own, which worked. Together, they also backed local talent.

There was a high school kid named Jay Springsteen, whose father and brothers liked bikes. When Jay was a kid he had a rare bone disease in his hips. He wasn't allowed to walk, so his dad cut down a mini-cycle and for two years, a motorcycle was Jay Springsteen's legs, literally.

In 1974, Springsteen was the hottest Junior anybody ever saw. He made Expert in time for the 1975 season, so Gawthrop and Klingsmith put the high school junior on a good XR-750.

The two different depths—Roberts and all his Yamahas against the XR-750 and all the good guys who built and rode them—made for a tight series. So, late in the year, Roberts pulled out the last of his stops.

That's more dramatic than it should be, maybe. Or there shouldn't have been such surprise and outrage.

When Class C was devised in 1934, there were motorcycle engines displacing more than 750 cc, and there were engines with more than two cylinders. These big machines were considered luxury craft, land yachts, so when the racing rules were drawn up the displacement limit was set at 750 cc and the other dimension, the number of cylinders, wasn't specified. There wasn't any need.

Later, as you've seen, triples and fours became sporting machines but although people certified them for Class C racing, and they took over road racing, they didn't work on dirt—not in 1969 and not in 1974.

But. Early in 1974, when the TZ750 became the ruling road-race engine, and there were no frame limits anymore, racer Steve Baker, tuner Bob Work and the guys at Champion Frames came up with a scheme to install the two-stroke four in a miler frame. They took the idea to Yamaha's management, which turned them down flat, so they went ahead on their own and built six of the devices.

Roberts got one. It sat in the back of his shop until late 1975, when he wasn't winning the points chase

Mark Brelsford always enjoyed winning races, but injury had him on the disabled list. Then he came back and was hurt again before he got back to speed, so he retired and went into ranching in Alaska. Harley-Davidson

A prototype XRTT at Daytona Beach, 1972. Visible here are the tachometer drive, from where the magneto used to be, the cutaway for the gear case cover, showing some modifications to the oil drain and return lines, and the indentations for the rider's knees against the 6 gal. fuel tank. And there's an oil filter tucked away below the rear carb. Petersen Publishing Company

and the Yamaha 750 twin wasn't winning on the mile. So . . . why not?

One reason could have been that it hadn't worked before. The TZ miler weighed 318 lb. and had 120 bhp, against the 300 lb. Yamaha twin and 290 lb. Lawwill-prepped XR, both with 85 bhp. But that didn't help in 1974. There were four other TZ milers entered for the Indianapolis Mile when Roberts arrived in 1975.

Against all that, Roberts wrestled his way into the main. At the beginning of the last lap, when Springsteen and Corky Keener were side by side on their Harleys, busy working out who was going to win, there was a terrible blast of noise and Roberts howled past, right on the edge, barely able to get around the final turn . . . and won.

Next time, at Syracuse, New York, there was no traction. At the second San Jose Mile that year, Roberts and the beast were sixth fastest in time trials and won their heat, but had a poor tire and finished sixth in the national.

At the end of the season, the AMA Rules Committee added the phrase, "maximum two cylinders" to the Class C rules. Not only did Harley-Davidson not object, Roberts didn't object. Scariest thing he ever rode, he said.

But, just as the TZ750 changed road racing forever, so did it change dirt track.

While that was going on, Gary Scott edged Roberts out of the national championship.

It was a hard loss to take. Roberts won six nationals and Scott won two. Yamaha had a wider range of equipment, but when the Harleys were strong, they were so strong and so numerous that they shut the other camps out of the points. So when Roberts won, Scott was close behind. It was a good, hard fight, with both economic and mechanical ramifications.

The racing engine that never was

Yamaha was catching up to Harley, so O'Brien scared some money out of the company, and the racing and engineering departments began work on a new machine.

It would have been a really new machine. O'Brien left the rules fairly loose, and the plan was to design an adaptable, almost modular motor. Jerry Long,

Here's a change of pace. These photos turned up in the basement of Pohlman Studios, the Milwaukee photography firm, in 1986. Their number sequence shows they were taken in late 1971. They are of a mockup of a design project, a 55 deg. V-twin with an overhead camshaft in each head and a single downdraft carburetor in the center of the vee. The frame is similar to the XR frame, especially at the steering head and the termination of the main backbone, but not identical; the center downtubes, the ones that should go from the rear of the tank to the swing arm pivot, aren't there. I would guess the ignition and alternator are behind the small silver plate on the black side cover. Harley-Davidson chief engineer at H-D then (1976), drew up a V-twin that could have been a 350/500 cc, or a 500/750 cc.

The included angle could have varied as well. In theory, the narrower the engine, the closer the engine can be to perfect balance, while the wider the angle, the more room for manifolds, ports and cooling air. Throughout the V-twin's history the angles were 42 deg. (Indian) and 47 (Vincent), 50 (Vincent), 52 (Honda), 55 (Kawasaki) and so forth.

The Harley design would have been, could have been, a 50 or 55 deg. included angle, with that vee picked mostly so there'd be plenty of room for two carburetors.

The 750 cc version would have had a shorter stroke and bigger bore than the XR engine, with shorter connecting rods and with one camshaft on each cylinder head with drive by chain, from a gear case on the right, like the XR.

Long, Zylstra and O'Brien got deep into the project, so much so that in the racing department it was known as Phase Three: what they'd do if the other guys got too close for comfort. But the engine was never made. There were some heads, cylinders, rods and pistons made, but a few castings and forgings was as far as the ohc 750 got.

Around 1970 or so there was a similar project, another ohc 750 that was dummied up just in case it was needed to combat the Triumph and BSA triples and the Honda 750/4. But that was stopped in its tracks, too, one guesses because AMF was willing to invest more in the plant than the product. Anyway, Triumph/BSA went out of business while Kawasaki topped Honda's 750/4 with a 900/4, a race in which H-D couldn't possibly have kept pace.

But, because nobody can manage to find pictures of the 1976 project, while some 1971 shots did turn up in the photo studio's basement, the 1971 750 represents the 1976 model here.

The Phase Three engine was never built because there was no clear and present danger. In 1975 Harley and Yamaha won all but two of the AMA Grand National races. The exceptions were two TTs, where lighter and more agile Triumphs in the hands of locals, triumphed.

For 1976, the British were long gone, out of business and racing. A poet could say winning the AMA title was the last thing they ever did. They probably would have gone bust anyway, but spending all that money didn't help nearly as much as an improved product would have. While the Japanese had no grasp of dirt track, the incredible TZ750 made road-race nationals not worth the effort. So Suzuki quit, Honda hadn't ever really been involved except for that one guerilla-style raid on Daytona in 1970, Kawasaki turned its racing bikes over to selected privateers... and Yamaha cut back too, releasing all its riders except Roberts.

This project would have made sense as a replacement for the Sportster, or perhaps as a more technically complex and advanced model to compete more directly against the British twins of the late sixties. But, as we know (and they didn't), the Japanese fours were on the way. Even so, this 750 twin with XR tank, disc brakes and electric start could have been a nifty road machine and racer. Harley-Davidson

Gary Scott, who won the national championship as a team member, then quit because he didn't think he'd be given as much support as he deserved. He came very close to winning the title again as an independent. Harley-Davidson

Meanwhile, over in Milwaukee, Gary Scott quit the team. He went away mad, taking the bikes O'Brien thought were owned by the team, and vowed to win the title again, for himself, riding Harleys, Yamahas and Triumphs, depending on the type of event.

Not taking sides, let's just say this was a clash of personalities. Scott wasn't plain selfish but he did put himself first and felt he should have gotten number one treatment from the company. O'Brien, a natural coach, put the team first and made sure that when Scott and Werner came up with something useful (which they did, as noted) the other team members were told about it and allowed to use it and even improve it; then the customers were let in on it. Both men truly believed their side of the story. It was a messy divorce and just as in real life, all an outsider can say is, sorry to hear that.

"Springer"

Jay Springsteen joined the Harley-Davidson factory team. Naturally. As O'Brien says, "If there's such a thing as a natural rider, Jay is it."

Along with Springsteen's incredible natural talent, and those formative childhood years spent riding when other kids ran, he may have been plain lucky because the new kid got Bill Werner as tuner. Teamwork counts, and Werner had already tuned his way into one national championship. He would go on to

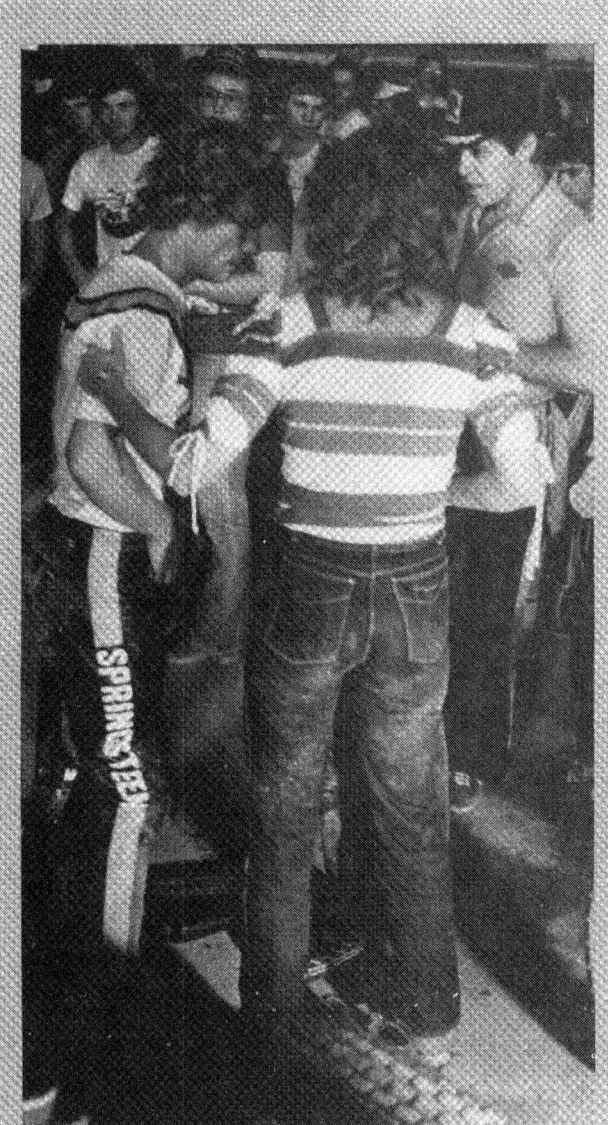

Photo by James F. Quinn.

The other side of fame

This is a picture a magazine editor would (and did) hesitate to publish.

It was taken July 3, 1982, at the Indianapolis Mile. A few minutes before this, Jay Springsteen was charging into the first turn, in mid-pack when the rider in front braked. Springsteen slammed into the other bike and was thrown to the ground at better than 100 mph. No broken bones, no blood—just a savage beating over every inch of his body.

Here, he's just been helped from the ambulance. His wife is guiding him to where he can sit down, and Bill Werner, right, isn't sure whether to help or not.

Look beyond that, at the sea of faces.

When the announcer said it was Springsteen who'd gone down, the crowd gave a collective gasp, then a sigh. As soon as he was brought into the pits, with no signal, no sign of any kind, everybody in the pits moved to the Harley space and formed a huge circle, just looking.

This wasn't the sick excitement mentioned only by people who've never been to a race. These watchers, like the people in the stands, were counting on Springer *not* to be hurt.

Springsteen is the risk-taker, the surrogate, the man who helps us to know we can overcome fear and boldly do what most people can't do. We want heroism and victory. We want champions who can by example make us all braver and faster than we really are.

But we don't want them hurt. We want them waving and smiling after narrow escapes; so the crowd surged into the circle because they had to see for themselves that Jay Springsteen was fine.

They meant well. But there he was—in pain and shock—in the middle of the crowd. Debbie and Bill got him to sit down and eased his boots off and then his leathers, got his jeans on, and Springsteen limped off into the dark, leaning on his wife's shoulder.

I was in the crowd, and I realized for the first time that there's another side to being the hero, the man on whom everybody else relies to perform another effortless miracle twenty-five or so weekends every year.

tune three more and become the winningest tuner in AMA history.

Technically, the next few years were straightforward stuff. It was factory policy to make XRs when the dealers needed them, and because the alloy version was everything the iron one wasn't, and a good Expert could make a living with an XR, there was another production run in 1976-77. The first alloy engine used the 42 mm intake and 35 mm exhaust valves from the iron engine, then went to 44/36 (or 35, depending on how you calculate the conversion between millimeter and inch) when the exhaust ports were improved, and finally when the XR1000 was designed, to 45 mm intakes.

The 1977 XRs, and the parts book, got exhaust pipes with built-in mufflers and megaphones designed by C. R. Axtell. There was a flap about noise making enemies for racing, and the AMA put in some new rules and tests, which led to abuse and then working out better ways to get the same power with less noise. (The factory thought it would be good to help the privateers with this.)

Jay Springsteen was a natural rider. He was also a natural star, possessed of that indefinable attraction that makes some people likable. He transmitted a sense of fun and daring, of pure untrammeled delight in what he was doing so clearly and powerfully that

Almost a team portrait, by luck. It's 1975 and Gary Scott, (64) is leading teammate Corky Keener and Jay Springsteen, the sensational rookie Expert on the Vista-Sheen XR-750, wearing leathers that look like team gear, although the official leathers were orange, and Vista-Sheen's were yellow. Behind Springsteen is a racer named Steve Drost, who rode Harleys some of the time but you can't be sure here. Trailing a bit is Rex Beauchamp (31), also on the team. Harley-Davidson

even the guys he'd just beat didn't mind losing, not to Springer.

The AMA had by this time been unable to ignore motocross, so there was a national series, and there were three kinds of pro license; dirt, road and MX. They didn't interchange, so Springsteen rode the lightweight road races on an RR-250 to earn points toward his road-race Expert ticket. By another stroke of luck, the H-D team wanted to run its 250 cc motocross bike in the big races, open class, so it made and certified a 341 cc version of the Aermacchi 250, giving the Grand National team the raw material for a short tracker. It didn't turn out to be a good one, but it did put him in the short-track nationals and gave him a shot at some points.

So there was lots of opportunity, wide open in some ways, as in Bultacos winning three of the five short-track nationals and Scott winning a half mile on his, um, the factory's former, XR and a TT on his Triumph. Roberts took two half miles on the 750 twin, a short track with the 360 two-stroke single and a road race on the 740/4.

Springer took four miles and three half miles with no versatility at all, just blinding speed and courage. The 1976 season came down to the last event, the Ascot half mile. Springer had to finish in front of Scott, that was all, and the plate was his. However, he dislocated a thumb in practice. Werner, who'd learned physical therapy when he wrestled in school, popped the joint back in place and Springsteen went out and raced for the win, which he got. Even O'Brien couldn't blame the kid for giving the fans what they'd paid for and not being careful about the points. Win the races, you might say, and the title will take care of itself.

Scott was a close second and Roberts an in-view third.

So for 1977 Roberts won all the road-race nationals *except* Daytona, which went to Yamaha's Steve Baker. The short-track rules had changed, so Springsteen won that race at Houston, then a TT, two miles and two half miles and was national champion again. This time he was followed by new H-D teammate Ted Boody, another Michigan rider, then Scott and Roberts.

Sounds much easier than it was, of course. But by the end of 1977, Kenny Roberts was the only man in the AMA's Grand National top ten who got there without a Harley-Davidson XR-750.

More about that. By the end of 1977, the XR-750 engine would deliver 90 bhp, maybe 95 if it was tweaked and wound up tight. Werner says they'd discovered by then that peak power wasn't always the best power, that the bikes went faster with power under the curve; for instance, good strong torque from 5000 or 6000 rpm on up. This wasn't always what the engineers and the guys back in the dyno room wanted to hear, so sometimes they'd race out at the old Milwaukee track, paved by then and no longer the place where proud Harley-Davidson racers came to have their pictures taken. These were blind races; the riders weren't supposed to know which engine was which, but they'd clock the laps, look at the identification and then go with the engine that did best on the track instead of on the test bed.

In the other camp, Yamaha Japan sent Yamaha US blank copies of the cylinder head. Then the Yamaha designers, tuner Shell Thuett and C. R. Axtell worked out the very best ports and valves. Their designs were incorporated into the cylinder heads which were made and approved as optional Yamaha parts.

The engine at first used what's called a 180 deg. crankshaft, with one throw up and the other down, side by side. That's the classic vertical twin. It has

A prototype of the mini-sump that cured the XR's problems with oil control, this is Bill Werner's collector box on the bottom of an XL engine case: Same design but the XL was cheaper to borrow than an XR case. Bill Werner

This is the entrance, the top, to the collector box, showing the baffles that guided the oil out of the crankcase and to the hose (not shown here) that took the oil back to the scavenge half of the pump. The bottom of the cases were milled flat and tapped for the bolts. The mini-sump was a running change for later (post 1974) XR engines. Bill Werner

good balance, because the two pistons and their rods and flywheels cancel each other out.

But when that version of the engine didn't whip the Harleys, the later versions—the full-race ones from Shell but carrying Yamaha identification code OU72 and cataloged as the XSDTC (XS for the XS-650 road model it was based on, DT for dirt track and C for model year 1976)—went to the other configuration for a vertical twin.

This is the 360 crank, as in 360 deg., with the two throws for the crankshaft next to each other, both up or both down at the same time. As one can easily imagine, this system doesn't balance nearly as well.

It has another sort of balance, though. The 180 crank, because the four-stroke twin has two power strokes every time the engine revolves twice, has a pattern of fire-fire-pause-pause. The 360 crank has a firing pattern of fire-pause-fire-pause. (If this explanation baffles you, pretend each hand is a piston and pump them up and down, first together, as with the 360 crank, then alternately, as with the 180 crank.)

Well, Ken Clark, Yamaha's racing think tank, believes now that the 180 version was too smooth and that even though the Yamaha 750 had 95 bhp in 1976 and 1977, when the best Harleys were straining at 90 to 95, the Yamaha didn't have traction. The 360 crank simply shook the engine apart. The team bikes and most of the private ones used Champion or Trackmaster frames. They had single shock rear suspension, good as anybody's. But the Yamaha didn't work as well as the Harley-Davidson and at the end of 1977, Yamaha officially withdrew from Grand National dirt track. As Harley had found on the road courses, it wasn't worth the effort. (Because the Yamaha had as much power as the Harley, and because Roberts surely wasn't at fault, and because Clark says it was lack of traction rather than horsepower, I wonder about the 45 deg. V-twin theory.)

At this point in our story, the good news for racing was that since 1974, the R. J. Reynolds Company, a manufacturer of cigarettes and related products, has backed the Grand National series with

The incomparable Jay Springsteen, at speed in the Santa Fe TT, 1977. Notable here is the oil filter atop the gear case, the use of a cast front wheel and spoked rear wheel, and the rock guard, apparently made from an old fender, on the front downtubes shielding the magneto. Harley-Davidson

hundreds of thousands of dollars and with promotional help in getting the public and press to pay more attention to the races.

This probably wasn't prompted entirely by disinterested sportsmanship. By no coincidence, the backing came right on the heels of Congress banning tobacco ads from the broadcasting media.

The bad news was that road racing and dirt track were almost completely separate, with, in effect, one factory dominating each side and nothing except the Camel Pro/Grand National/AMA championship tying them together.

But nature abhors a vacuum, bless her heart, so for 1978 Springsteen didn't have Kenny Roberts to race because Roberts reckoned he'd do better against the fabled GP crowd, the so-called world, than he'd do against Springsteen and the Harley-Davidson XR-750. And he did. Roberts paused only to win the Daytona 200, at last, before going to Europe and teaching the effete foreigners how cowboys ride motorcycles.

Meanwhile, there was a new sub-era at home. Springsteen won his third Grand National title in 1978. It was another tough one, right down to the wire. The main opponent was a privateer, Steve Eklund, a northern Californian tuned and backed by a mathematics professor (no kidding). They were outsiders, but because they worked and rode hard and because all the parts they needed to go fast were available from Harley-Davidson or the aftermarket, they could mount an attack that lasted until the last

The 1977 Harley-Davidson Team portrait. From left, Ted Boody, Jay Springsteen and Corky Keener. This bike illustrates the changes made since the XR began production in 1972: Squint closely and you'll see that the engine has been moved forward and the front lowered, while the steering head is a bit farther ahead of the tank. This is the TT version, so it has disc brakes, and as you can barely see, a mini-sump with hose just above Springer's right foot. Harley-Davidson

race of the year. Once more, Springsteen finished first in the race and thus the series. Eklund was second in both.

The independents, or privateers, were back. Most of them rode Harleys, true, but the point here is that out of twenty-eight nationals, twelve winners were on Harley, BSA, Yamaha, Triumph and Norton teams.

There were at least eight capable design and construction outfits making good frames for the XR engine. There were alternate sources for tanks and seats, for forks and shocks and even three brands of certified racing tire. Lawwill had retired from riding and had become the leading supplier of cams, ignitions and other parts. He also built engines and/or turned other people's engines.

People didn't seem to mind, either, that the race would be this guy on an XR versus that guy on an XR. They came for good racing, and as reporter Pete Lyons put it, good racing is when you don't know who's going to win.

But then Springsteen suddenly evened the odds.

He got sick. He'd be fine until just before the race, when the seemingly relaxed Springer would be knocked off his feet, literally, with stomach spasms. He couldn't eat, and when he did he couldn't keep it down.

Jay Springsteen was a more complicated man than people had thought. The outward part, the easygoing, relaxed, smiling man who liked the outdoors best, was the first level. Then came the second level, the showman who revelled in the roar of the crowd.

The third level was an intensely private man, at ease only with family and a few close friends, awkward with strangers and the press and distrustful of

Springsteen had his Expert road racing license by 1978, when he ran the road-race XR on the shorter tracks, such as Louden here, in case he could pick up a few points. He had a good time and entertained the fans, but mostly, the bike broke. *Cycle World*

the swarms of slick talkers ready to help him get what *they* wanted.

But Springsteen really liked the fans and always had time for them. What he wasn't prepared for was their expectations. The crowd always counted on him to do their risk taking, to defy the odds and come through victoriously. He didn't know how to deal with that kind of pressure.

Springsteen lost to Eklund in 1979 and he went through batteries of tests. The doctors proved only that if you took ten guys, same age, height, weight and muscle tone as Springsteen, and lined them all up for a 100 yard dash, Springer would win. Every time. He'd win because he would *want* to win more than anybody else in the race.

What the doctors didn't do was find a cure. So with careful diet, exercise and pacing himself, Springsteen has since managed to be a contender and the winner of more AMA nationals, forty at this writing, than anybody else in AMA history. But he was never again the overwhelming favorite that he was in 1976 and 1977.

What this did (disallowing Werner's remark that Springsteen's being sick was nature's way of giving the other guys a chance) was, well—give the other guys a chance.

Randy Goss

There were twenty-six nationals in 1979, with fourteen different winners, and Eklund the series champ. In 1980, new H-D team member Randy Goss—yes, from Michigan—beat Hank Scott (Gary's brother) by one point in the last race of the year, and the factory got the plate back from its customers again.

Goss was another low-key rider, a former protege of Bart Markel's. He married into a racing family: wife Vicky was an ice race winner and her mom made leather suits for racers. When Vicky was pregnant, those who asked when the baby was due were told, "between Santa Fe and Indy"; Vicky simply based her

Springsteen's natural home was the mile, where he was the only rider who could run wide open into the turn and pitch the bike sideways at 120 without shutting off on the throttle.

calendar on racing, and why not? Goss and Vicky and tuner Brent Thompson, who'd learned his trade on the two-stroke road racing and later the motocross projects, were a closely aligned partnership and won on grit more often than brilliance. But Goss was always in the points and they added up, reversing the old saying to "take care of the points and the wins will take care of themselves."

A fading star

The 1980 season saw the beginning of an end, if not *the* end: The factory built 200 more XR-750s, the last production run for the model. They were the latest step in an evolution that began generations before Harley-Davidson adopted the word.

The 1980 models were in the TT mode: The frame came with the steering head angle at 24 deg., two or three degrees steeper than had been specified earlier. The front of the engine was one inch lower in the frame; the rear of the frame was higher because the rear shocks were 14 in. long where they used to be 12.9 in. or shorter. The exhaust system was a 2-into-1, sized in diameter and length for extra punch off the turns, power that can be used best in TT races, instead of the top end power needed for the mile.

In keeping with the XR tradition, the frames were actually made by Champion, an outside supplier, to

Modified TT frame used threaded mounts and clevis for the top of the shock because Werner wanted to vary static ride height of the bike, but didn't want to vary the spring's preload at the same time, which is what the standard system does. (Sorry about the tape; Werner didn't know this frame would be dragged out for inspection.)

specifications arrived at by the team engineers and tuners and riders.

The factory's official figures say there were 543 XR-750s made and sold by the factory. That's an exact-sounding total, better than the 500 to 1,000 estimate for the KRs and KRTTs made between 1952 and 1968.

But at the same time, it isn't guaranteed. For one thing, although the factory hasn't made complete XRs

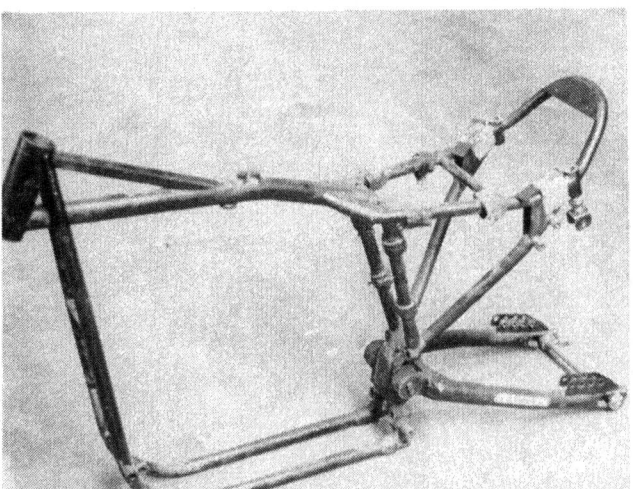

Straight from a dusty corner of Harley-Davidson's secret racing shop, a bit of technical history: a frame Bill Werner modified for Jay Springsteen's TT bike. It began as a normal XR frame, with stock backbone and steering head, but notice the rear hoop bridging the rear top tubes and shock mounts, with the extra brace in the middle of the rectangle and the gusset at the inside of the hoop, the gussets in front of the shock mounts. Werner says the frame can be banged out of alignment if dropped, unless the two rear extensions are joined into one. And, note the four sets of mounting holes on the swing arm bracket for the shock. Moving the shock in its mount changes the leverage and the effective spring and damping rates of the shocks.

Even more subtle, this is the swing arm pivot. Werner has made the solid mount, the twin tomahawk, into an adjustable pivot. Leverage, thus weight transfer, thus traction can vary with track and engine and so forth, so an adjustable pivot height can improve on a setting that's otherwise a compromise.

since 1980, it does stock and sell the engine cases. (All Harley engines, even racing ones except for the bought-out 500-R, are made by the production department, so they can keep their precision hand in, and still control quality.)

All the parts—the flywheels, covers, gears and so forth—are in the catalog and stocked in the race department if not the local dealership, so you could right this minute buy a complete set of parts, put them together and presto, a new XR-750! Moreover, there are a handful of dealers around the country who bought XRs in 1972, 1977 or 1980 and parked the machines in the showroom. If the money was right and circumstances were in your favor, you could buy a certified new XR.

Furthermore, there's an accounting question. If a tuner and rider form a partnership and one supplies the engine and the other the frame and running gear, is that another XR for the books? Or is it just two halves?

And, if (as happened) Carroll Resweber takes the remains of the XRTT that Mark Brelsford crashed at Daytona in 1973, and redoes virtually everything in the frame while fitting an entirely different engine, suspension, brakes, tank, seat and fairing, and the team calls it *Lucifer's Hammer*, the XR1000 road racer, is it? Or is it Brelsford's old XR-750? Or both?

Or does it matter? The factory built as many XR-750s as the professional racers needed and would buy. Can't ask for more than that.

Renewed competition

Meanwhile, as Harley re-filled the racer pipeline, Honda and Yamaha returned to the Grand National wars. Just why, nobody said, but in 1981 both of the former rivals returned. And although the rules had loosened so you didn't need a production replica, they both had exactly that.

Perhaps it was the marketing: Yamaha had introduced a 750 cc V-twin; 70 deg. included angle, two valves per, air cooled and shaft driven. They used that as the basis for a track machine, converting to chain drive and using a custom-designed (*in* the United States *for* the United States) frame with single shock rear suspension.

For corporate reasons, Yamaha didn't actually come right out and sponsor this effort. Instead it hired a contractor, Roberts-Lawwill Racing, to develop and campaign the bikes. Yes, that Roberts and that Lawwill. Yamaha racing director Ken Clark was in on

Springsteen (9) and fellow former national champion Steve Eklund on a Yamaha 500 at the Houston TT, 1981. Springer is shown on the last version of the XR, the TT model, so called because it was taller and had more weight in front. The forks were Marzocchis, replacements for the Cerianis of the seventies, and the fuel tank was made of aluminum rather than fiberglass. *Cycle News*

the deal, though, with former H-D team rider Corky Keener serving as liaison man. The main rider was Mike Kidd, the third member of the rookie class of 1972 and a solid racer who also could organize well: He'd persuaded the US Army to sponsor him on his own Harley-equipped team. Junior rider was Jim Filice, an upcoming new expert from northern California.

Honda seemed to have gotten deeply into Class C for the first time in large measure because the AMA Grand National title was the only championship in the world Honda hadn't won yet. Its entry began life as a road bike, a 500 cc 80 deg. V-twin, water cooled, four-valve, shaft drive ... and with the cylinders crossways in the frame, not fore and aft.

Honda back home turned this engine a quarter turn, so the cylinders were fore and aft, chopped off the shaft drive, installed a sprocket and bored the thing to 750 cc. The company hired Jerry Griffith, a private tuner who wasn't very well known, to run the team and assigned Freddy Spencer, the scrupulously polite road racing star, to be number one rider and borrowed privateers for when Spencer was busy in Europe.

Summing up a vastly entertaining season, people had forgotten that Spencer was fearless and tough, and that in his early childhood he rode on the gritty tracks of backwoods Louisiana and Texas.

Now, a man can be made of iron and not take the Lord's name in vain, but, this machine was terrible. To watch Spencer all but stand on the front axle, wrestling the goatlike motorcycle and failing to keep it aligned with the track, made you want to take off your hat and burst into giggles at the same time. (But don't feel badly, the Big Red Worm is going to turn with a vengeance.)

Over in the Yamaha camp, things were different, and better. At first, anyway. The V-twin was beautifully prepared and turned out. If it didn't go as fast as the Harleys—well, it was still early.

One of the better-than-fringe benefits of a factory ride is having resources like this: Bill Werner and Jay Springsteen go to the races with two bikes, both fresh but usually one tuned for torque and the other for peak, or with slightly different chassis settings. Springer tries them both and rides the one he likes best. Here, at the Indianapolis Mile, 1982, Werner is trimming the tire tread while Jay keeps the bike from falling off the stand (joke). The track was very wet, so the tires have maximum tread allowed and Werner is making the edges as sharp as possible. Notice here the addition of CD ignition, with the housing on the gear case cover and the battery in front of that. James F. Quinn/*Cycle World*

Then it wasn't so early, and the Yamaha still didn't work. Nobody knew why. The team got new frames; didn't work. They moved the engine up and down, back and forth. They raised the swing arm pivot. They lowered the swing arm pivot. No good.

For Honda, same sort of thing. The bikes had different exhaust pipes and different suspension, but while Spencer was taking road races away from Yamaha, and private Honda singles and Yamaha singles did fine, the V-twins didn't.

But Mert Lawwill built Harley racing stuff. He had a couple of XRs around the shop and, well, everyone got racing fever. And because Kidd had been in the points, Filice did the testing for Yamaha, and Kidd spent the last half of the season on the Harley, and won the series title. Did it on points; one of those mathematic jobs where Goss won the last race but lost the plate because Kidd was second.

A few sidelights: One, if Kidd hadn't won, the title would have gone to Gary Scott, who was not popular with Harley. The H-D team in fact loaned Kidd, the Yamaha-backed rider, a fresh and perfect tire for the last race. That could have made the difference. Two, after the event, the T-shirt folks came out with one that said, "HARLEY-DAVIDSON: First Choice of the Yamaha Racing Team." Three, when Yamaha didn't renew the contract with Roberts-Lawwill, Mike Kidd and his No. 1 plate were hired by—who else?—Honda. So this time the wags said, "Harley Won It, Yamaha Paid For It, Honda Bought It."

But that was mean and cruel (tee, hee, hee).

Technically, neither new machine worked. As an illustration of how far tuners go, back in the days of the English twins with 360 deg. crankshafts, somebody worked out that you could get maximum thud, so to speak, if you rearranged the camshaft and ignition timing and fired the two cylinders at the same time, a double boom followed by the three rests. This would be a twin that fires like a single, known in the sport as a Twingle. Yamaha and Honda both tried this system, to no avail.

Bill Werner also used it on one of Springsteen's TT bikes, and it worked; at least Springsteen won, which might not prove anything. Years later, Werner recalled that he'd discussed the idea of Twingles with Jerry Griffith, the Honda tuner, but he (Werner) didn't tell Griffith everything there was to know about the trick.

Another important point in all this was that both the Yamaha and the Honda were wider than the Harley in the vee. And, at one point another tuner showed up with a 90 deg. Ducati V-twin in a dirt frame and with a ranked Expert rider, but they didn't make the main event. So there's no proof that the narrow vee is the secret, but the wide ones didn't impress anyone.

New faces and new bikes

Ricky Graham, a northern Californian, had always been fast, but he was erratic and crashed a lot. His dad helped, then Graham got some good sponsors and tuning, but the combinations never clicked.

Then he met Tex Peel, who came (you don't need to be told) from Texas but worked as a skilled craftsman for Buick in Flint, Michigan. The oversized Peel and the easygoing Graham hit it off. Better, they worked together well and hard, and they understood each other.

Graham began the 1982 season by doing something nobody had ever done—winning the Houston TT with an XR-750. It was as smooth and effortless-looking a ride as anybody ever did, never mind that Graham's cracked ribs were taped from a practice crash, or that the bike was a 1972 frame modified back in 1975 by the Vista-Sheen team. Peel had a

Mike Kidd, who parlayed a private Harley ride into a national championship, won on a Harley with Yamaha money, then got a Honda factory contract. After that, he retired to begin a successful career as—what else?—a race promoter. *Cycle World*

Lucifer's Hammer, Battle of the Twins champion

As he swung into the exit of the infield's U-turn during practice for the Battle of the Twins (BOT) championship race at Daytona Beach, 1983, Jay Springsteen heard an annoying little noise: It was Jimmy Adamo, on the reigning-champion Ducati, putting his bike's front wheel just inside Springsteen's rear wheel.

Humph! snorted Springer with a gesture visible all the way across the infield, thinks he can do that, does he? So Springsteen ostentaciously whacked the throttle full open, the Harley's front wheel just barely grazed the surface of the track and away they went. Next day in the race, Adamo never got close enough to get Springsteen's attention.

Perched on a high place in the infield, holding his breath for each and every lap, was Louis Netz, who was then second in command (after Willie G. Davidson) at the H-D design studio. Netz on his own time is a tuner. Harley doesn't pay for vacations, so Netz was officially and privately in Daytona Beach tuning Randy Goss' dirt trackers for the dirt-track series held during Cycle Week.

But of course he came to the road races and did what he could, and when the flag (finally) fell, Netz leaped up and sprinted to the pits to get on the phone and let all the folks back home know *they had done it.*

It wasn't easy. By 1983, Harley-Davidson had been out of road racing, in the sense of winning a national, since 1972. Road racing had been the province of Yamaha, with challenges from the two-stroke multis of Kawasaki and Suzuki, since then.

But the fans had objected and so the AMA made some changes. They created a class called Superbike, which was a closely controlled and inspected series for production-based road bikes that were in large measure stock—souped for racing but still obviously related to what the fans rode to the track.

As can be seen here, *Lucifer's Hammer* **used a semi-dry sump, the long, thin component beneath the engine, which made draining the crankcase and rocker boxes a lot easier and kept the weight down. That's the old KRTT-style seat and fairing, but the front wheel was 16 in. in diameter and the brakes were triple disc. The frame came from the back of the shop; most of it was crashed at Daytona in 1973. But this machine, sponsored by the factory-backed Harley Owner's Group, has won the Battle of the Twins title three years running. Rider is Gene Church, a former dirt expert.** *Harley-Davidson*

And the purists got a more closely controlled series. It was for four-stroke twins, 1000 cc limit, but with classes for stock and for smaller engines. They called it Battle of the Twins. It took hold in the United States and in England, the natural home of all those Nortons, Triumphs, Ducatis and Moto Guzzis.

Then, in 1983, Harley-Davidson took most of the Sportster XLX, the cheapest bike in the line, and topped it literally with the good parts from the XR-750. It was called the XR1000. (Why they used a hyphen for XR-750, as they did in the parts book, and not for the XR1000, nobody knows.)

The XR1000 was a fast road bike, with nothing extra added on and with the alloy heads and dual carbs from the XR, but the iron barrels, bore and stroke of the XL.

The BOT rules allowed racing versions with production engines, so the racing shop was given the assignment of building a racer, as well as designing and testing for the customer's road-legal version.

This is either a tribute to sound engineering or evidence that H-D had been out of road racing too long. H-D began with the frame from the XRTT wrecked by Mark Brelsford at Daytona in 1973, right, ten years earlier.

Because Cal Rayborn had been the best road racer last time the team had looked, and because O'Brien had Rayborn's old bike tucked away in his own garage, and because the current crop of Harley racers (read here Springsteen) wouldn't have any validated ideas of their own, O'Brien and Resweber built the BOT racer out of

The Buell RR1000. Designed and built by former-Harley engineer Erik Buell, this machine had a space frame (many small tubes) and was lighter and tighter than *Lucifer's Hammer* or the XR-750 it was built from. H-D and HOG planned to build *Lucifer II* from a Buell, for BOT in 1987 and maybe Superbike after that—rules, economy and the AMA willing. Harley-Davidson

Brelsford's old frame, with the various distances and angles based on what Rayborn used way back when.

Resweber added two tubes to the top of the loop, the engine cradle, and beefed the steering head and swing arm pivot and made a new, stiffer swing arm. Wheels were 18 in. castings, with a 6½ in. rim on the back and a 4

Gene Church and *Lucifer's Hammer* at Daytona Beach. Frame and suspension weren't the latest things, but the engine had so much power, always on tap, that the bike was competitive years longer than it should have been. Harley-Davidson

Buell's frame was much stiffer than tradition says is optimum for handling. It used a series of rubber mounts and links to isolate engine vibration from the frame: one such linkage can be seen to the left of the cross brace triangulating the top frame tubes.

The Buell was shorter and narrower than the Harley XR1000, so the front carb had been angled out and the rider's right knee went behind it, rather than angling the carb back and having the rider's leg outboard of it. Scott Darough/*Cycle*

in. rim in front, and with suspension from Marzocchi in front, Moto-X Fox in back.

The oil system is one of the team's favorite tricks. The bike got a huge sump, an oil tank fabricated from aluminum sheet and mounted below the crankcases; a semi-wet sump, you could say. This was obviously easier to drain into. Also, the long, thin aluminum tank was in the airstream for cooling and lowered the center of gravity while being tucked out of the way, so the other bits didn't need to be compromised.

For drama, O'Brien called the beast *Lucifer's Hammer*, out of Irish mythology, he said, even though nobody else had heard the story, grandmither from the auld sod or no.

For instance, O'Brien assigned Springsteen to ride the bike's first race. It was routine for those who'd seen Springer in action, as he drifted both wheels and steered with both wheels, and generally carried off all the speed secrets experienced road racers said couldn't be done. (He did the same thing on the dirt in 1975 and in the Daytona Superbike race in 1986.)

At any rate, Springsteen won *Lucifer's Hammer's* maiden voyage and the bike was transferred to outside, with Don Tilley, a North Carolina dealer and tuner doing that job and the riding by Gene Church, a dirt expert recruited because he was good and brave and had no bad habits to *unlearn*.

The Buell RR1000, shown here fitted with an XR1000 engine, was very obviously a lot different from the Sportster or an XR under their respective skins. The frame was straight tubing, a collection of triangles that formed a bridge between steering head and swing arm pivot: no massive this, no fabricated or cast that. This example has electric start—note the starting housing portion of the primary drive cover at top right of the engine—with a 12-volt battery and oil tank below the seat. Scott Darough/*Cycle*

The Buell swing arm pivot was inflexibly and neatly made an integral portion of the frame, bracketed and braced by the four rear downtubes. The single rear shock mounted beneath the engine, the only place there was room for it. Scott Darough/*Cycle*

The combination won the Grand Prix class—the biggie—in the BOT series in 1983, 1984, 1985 and 1986.

Not without incident or effort, and not by winning every race. Adamo and the Ducati were worthy opponents, as were visitors like Australian Paul Lewis and the Cosworth twin, who proved as fast and sometimes even more durable, never mind modern.

The *Hammer*, though, benefited from subtle little differences.

For one thing, O'Brien had retired, by no coincidence, to a lakeside cottage with just enough room for him, his novelist wife and their giant but well-trained dog, and was within easy reach of Tilley's shop. O'Brien signed on to run the engine department for the Skoal Bandit NASCAR team; racing engines are racing engines, he says. And for fun, O'Brien has been liaison to the old racing shop, the sponsoring Harley Owner's Group (HOG), Tilley and a host of outside helpers.

The ancient, outmoded foundation for the *Hammer* has held up better than anybody could hope. The engine has gotten better and better. In theory, it's an XR1000 and has the 81x96.8 mm bore and stroke to prove it. But the iron barrels have been artfully anchored in the cases and the cases are XR—stronger material and built to use the better bearings, the needles and superblend bearings developed for the XR.

In his guise of godfather to this team, O'Brien provided his unique brand of help. First, there was an extraordinary pair of cylinder heads, with different ports and titanium valves, carved out of castings made in 1982, two full years before the XR1000 and its racing descendent were built.

"How'd you happen to have them?" I asked.

"I had them made. I was planning on stuff like this. I used to have parts made in advance, that wouldn't get used for years, but I wanted to have it there. Sooner or later, it was going to be needed," O'Brien said.

Second, O'Brien had an unequaled depth of resource, in the NASCAR team and its connections, and the thirty-some years of friendships built from motorcycle racing

Gene Church in action. *Lucifer's Hammer*, like the dirt-track XR, was smaller than its engine and performance leads one to expect, and as the cornering angle here illustrates, the machine and the rider have adapted well to the pavement. To keep in balance, a motorcycle must lean in direct ratio to its sideways cornering force, so one lateral g of grip allows 45 deg. of lean; judging from this shot, the *Hammer* generated more than a g.

on all sides of the fence. And all this time, he's stored away everything that's been done with this engine.

When it first appeared, the *Hammer* had 107 bhp. Because O'Brien used to average one bhp per year with the old KR, if he's done that well, the XR1000 now has 110 bhp. If he's done better, more likely, it has 115 or so. And most of the time it finishes the race.

Lucifer's Hammer has gone to a 16 in. wheel in front to use today's best racing slicks and sharpen the steering. The brakes are Brembo discs, the seat and fairing are good ol' XRTT. And they look it. O'Brien says nobody's had the nerve to weigh the machine, but 385 lb. would be an educated guess on the high side, while one published test said 350. Maybe.

This racing machine is as old as it looks. Say, twenty years for the overall design, fifteen years for the frame and engine, and still champion.

Today, early in 1987, HOG, Don Tilley and Dick O'Brien are working on some plans. A former Harley engineer named Erik Buell, one of those backyard geniuses, designed and began to build a motorcycle seemingly perfect for BOT and/or Superbike.

The Buell RR1000 will accommodate the XR-750 engine, the XR1000, or the new Evolution Sportster engine that's based on both the earlier powerplants and probably has more potential than the other two. The Buell is enclosed and is right up there with bikes from, ahem, other continents in style. It's smaller and lighter and, my gosh, it had *better* be a better frame, unless we've all wasted twenty years.

So. What if HOG comes through and puts Springsteen or Parker on one of those machines for BOT? What if Buell builds enough to qualify as a Superbike? And what if the AMA got hold of itself and worked out some rules that would have an equivalency formula for two-strokes and four-strokes, 1100 cc twins against 75 cc fours?

Yes, indeed. If we did it right, we could reinvent Class C and the Production Formula, and we could have another fifty wonderful years of racing.

This Harley-Davidson chapter closes. But the story goes on.

A practical mixture of old and new, *Lucifer's Hammer* **obviously had the seat and tank derived from the old KRTT, along with a 16 in. front wheel and slick tire, as seen in Grand Prix racing. Also visible here are the rearset linkage for the shift shaft coming out of the case's left side, and the oil sump, the long, thin fabrication below the cases.**

reputation as an engine genius, while he himself said it was all due to Ron Alexander, the tuner who'd done all the work on the iron XR-750 and early alloy engines, and who'd been transferred back East.

Right, the old hobby, with the outside guys teaching the factory tricks. Except, they had inside help; remember Phase Three, O'Brien's never-built ohc 750 racing engine? O'Brien likes short-rod engines, so does Peel and so does Alexander, and somehow the rods made for that stillborn short-stroke, short-rod ohc 750 turned up in Peel's best XR-750 engine.

It was another classic, right down to the wire for the Ascot half mile, the last race in the season. Crafty Randy Goss figured out the track surface and won, while Graham was sixth and Springsteen eighth in the race. The year's finish was Graham, Springsteen, Goss. Filice on the Yamaha didn't win any races; Kidd on the Honda didn't win any races and sat out the last part of the season with a broken leg. The year's new face was a soft-spoken (if that's possible) Texan named Bubba Shobert. He won two nationals, one on his own Harley and one on a bike he borrowed from Peel, leading Peel to quip that he'd have to be more careful about who borrowed his bikes.

Yamaha retired for 1983, Honda kept on, while Peel and Graham learned, to their sorrow, that winning the title was possible if you knew how to ride and tune and prepare, but cashing in on the title when all you knew was racing was something else and something at which they weren't nearly as good.

Times were getting tougher. All the racers' costs were going up, inflation and so forth, while the purses weren't. To put on a complete, private program with bikes, spares, van, gas and food would take about

"Who me?" says reigning national champ Randy Goss; "Yes, you," gestures former national champion Gary Scott, who said Goss' leaking oil tank had made the track impossible to see. And from the look of the bike, he's right. Scott is mad mostly because Goss managed not to see that black flag until he'd won the heat. It's always better to win first, argue later. Steve Kimball/*Cycle World*

$50,000 for the season. Purses couldn't cover that. Nor would or could most sponsors. Thus, a tier system developed, with the factory-paid guys on the top, then the riders with full and adequate sponsorship, a program as they call it, then the part-time experts, the local guys who came to the races they could afford, backed by Mom and Dad or the local dealership (unless Mom and Dad *were* the local dealership).

The Camel Pro points fund helped a lot. What helped more was paying the fund in two parts, one at mid-season and the other at the close of the year. If it hadn't been for the midyear payout, a surprising number of private racers couldn't have paid for gas for the second half of the season.

Then, at midyear 1983 Honda introduced its second series machine. Funny outfit, Honda. The guys never apologize and never quit. While they'd been racing the NS750, the one based on the old CX500, they'd introduced a new V-twin, a fore-and-aft one, in 500 and 750 cc sizes. And they'd bought an XR-750 engine, taken it home to Japan and taken careful notes on it. They hired Gene Romero, the former No. 1 for Triumph, and Rob Muzzy, the superbike tuner laid off when Kawasaki could no longer afford to race against Honda in that league.

For the spring Ascot meet, Team Honda unveiled the RS750. It was an ohc four-valve V-twin with the cylinders at 45 deg., but with the crankpins staggered so they were at 90 deg. to each other; the firing order was wider than the Harley's. And with two crankpins side by side instead of one pin in knife-and-fork rods, the Honda engine was wider than the Harley, while having more room for ports and pipes and manifolds.

The new Honda engine was a copy of the old Harley engine. Honda might have taken the best of the Harley design and improved on the worst, but that's been part of racing since the second man noticed the first man had an open exhaust.

Nor did the Honda win, right off. The 1983 season was the last all-Harley year, you could say, with Randy Goss coming through on points. Filice switched to Harley and won two races, Shobert won two and Gra-

Ricky Graham, on Tex Peel's circa 1972 XR, en route to the Harley 750's first Houston TT win, in 1982. Peel said Graham would win the national championship with his consistency and speed on the miles, and that's just what he did. David Edwards/*Cycle World*

ham racked up the last three of the season, including the Pontiac, Michigan, TT in which he lapped half the field and won going away . . . except that if Goss got any points, he got the title. Goss made the final, and thus got the points and the title.

Then Honda hired two new riders: Ricky Graham and Bubba Shobert. Again, that's fair in racing. They didn't have rides, there wasn't room on the Harley team, and as any racer or athlete will (or should) tell you: You have a finite time for your career, take advantage while you can.

Changes come quickly

But 1983 was the last year the XR-750 ruled Grand National dirt-track racing. It was close when Honda had the power and Harley had the numbers, but Honda's edge on power meant more racers shifted to Honda. Graham won the title for Honda in 1984, followed by Shobert, then Goss and then Scott Parker, the newest member of the Harley team.

Things happened faster and faster. Shobert won the national championship in 1985. The title chase and points race concentrated. Goss was national champ by six points in 1983, Graham by one point in 1984, Shobert by 56 (!) in 1985, Shobert again by 96 (!) in 1986. Shobert had the plate welded to his Honda, as it were, three races before the end of the season. At the end of 1986, Shobert held the AMA record for career mile wins (19), mile wins in a season (6) and overall wins in a season (9).

Meanwhile, things happened as quickly in the real world. O'Brien managed to outwit mandatory retirement by one year, but that was all. His assistant and replacement, Clyde Denzer, was laid off. So was Resweber. The racing shop was closed, the tuners transferred to other jobs. The team dissolved. Springsteen and Parker were given money at the beginning of the season and were supposed to hire their own tuners, pay their expenses and keep the prize money.

The burden in 1985 and 1986 fell on Scott Parker and Chris Carr, another new rider, this time from California. Carr had a different deal, in that he was first hired by and rode for Ron Wood, the Rotax distributor for the United States and a top tuner and talent scout. Then Carr had two sponsors, Wood for short track and TT, on the Rotax 500, and Mert Lawwill for miles and half miles, on Lawwill's superfast XR-750.

Adapting to survive

On still another front, the AMA either caved in or faced reality, by splitting the national championship into two titles, dirt and road. The way one reacts to that depends on whether you appreciate the hard

Reigning champ Graham and three-time champ Springsteen, at San Jose 1983. Peel built and Graham rode, sometimes with more energy than control; they broke or crashed when they didn't win. But Springsteen and Graham and their machines were well matched and ran like this for two seasons. Steve Kimball/*Cycle World*

and fast and close racing of the uniquely American system, or if you think the two forms of the sport are so different as to be incompatible. (The third view, that the more championships you have the more they can be marketed and exploited, is held only by vile commercial interests. If you feel that way, please put down this book immediately.)

If this all sounds as if an era is drawing to a close, it was doing exactly that.

For the other side, some disquieting facts: Grand National racing didn't have the public appeal in 1986 that it had in, say, 1966 or 1976. At one time, a double bill at the Astrodome drew 75,000 fans. By 1985 or 1986, everyone would have been happy to see half that. At Anaheim Stadium, 70,000 people packed in to watch Supercross. At Ascot, racers going twice as fast, on the edge with three times the horsepower, drew 7,000.

But the real kicker, the new meaning of the old cliche about good news and bad news, was this: Harley-Davidson's management bought back the company, with a lot of help from banks. Then they went public again and sold stock, after having put all the bank money and their own money into the product. They began building the best Harleys ever. They came out with the Evolution big twin with vibration-isolating engine mounts and five-speed transmissions and a bare-bones Sportster, and then an Evolution version of the bare-bones Sportster.

Good versus bad? Yes, because in 1984, 1985 and 1986, when Harley-Davidson was losing the national championship, and when the Hondas outran the Har-

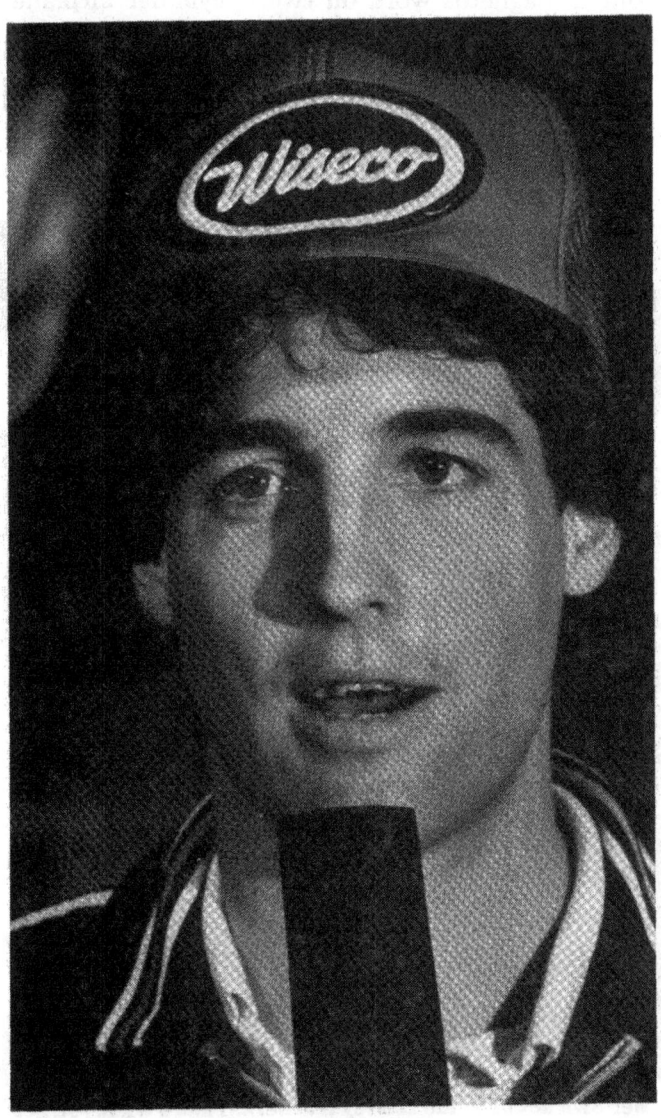

Ricky Graham knew in an interview before the race that he had to win—which he did—while Randy Goss had to finish out of the points. James F. Quinn/*Cycle World*

Tex Peel and Ricky Graham, who found it easier to win the championship than to keep it or make a profit from it, at the Pontiac TT, last race of 1983. James F. Quinn/*Cycle World*

leys, Honda sales (and motorcycle sales) went into a slump.

Harley-Davidson sales didn't. The FXRs and the Softails went at full pop, and 10,000 Evolution Sportsters sold as fast as the factory could build them.

Lessons here were: One, Harley fans won't go to the track to watch their brand lose, while Honda fans won't go to the track even if their brand wins.

Two, building good motorcycles that sell in record numbers is a better way to keep your company going than spending the money on racing. Sorry to admit that, but the figures are all there and that's what they show.

How the pieces fit

We aren't done yet. As this era draws to a close, the XR-750 is still a viable racing machine, still winning and still the first choice of the new professional, albeit that's largely because the XR can be bought as a good, solid, used ready-to-race motorcycle while the Honda RS750 costs twice as much (in 1987) to buy and assemble. (Honda sells the engine and C&J sells the frame, which it builds to Honda's design, the same way Harley does.)

Politics and people aside, the XR-750 is everyman's racer, a kit bike with established form and techniques and, yes, some secrets as to how to make them work, how to make yours faster than mine.

So, some looks at that.

Ignition

This is the last frontier, the final factor in making the thing work, as opposed to making it work better or best.

At first glance, that shouldn't be so. When the WR was introduced in 1941, one way you could tell it was a racer was the magneto ignition. Same for the KR, XLCH and XR, and for racing cars and airplanes for three or even four decades. Magnetos—self-contained, self-powered ignitions—have been the logical choice. And if magnetos work on twelve-cylinder airplane engines and eight- and four-cylinder car engines, why, surely they are the best and most reliable ignition for a plain little V-twin?

No. Bill Werner used to have two ignitions on Springer's bikes, one magneto and one coils and dis-

Post-race at Pontiac 1983, from right to left, announcer Larry Maiers, new national champion Randy Goss, the race's promoter, and Goss's wife Vicky and daughter Janice, who is wearing a surprise present from Grandma, her own set of Team Harley leathers. That's what you get when Grandma's business is . . . making racing leathers! James F. Quinn/*Cycle World*

tributor. They were wired in series, that is, if one went sour you could swap a few wires and run on the other. Then he went to two points and coil systems wired in parallel, both firing all the time, to dual-plug heads. And in 1984 he designed, built and began selling an electronic ignition, with generating coils and CD ignition triggers, running off the front of the gear case where the magneto used to be.

Why? Because in one earlier season he had fifteen, yes fifteen, magneto failures.

Pieter Zylstra says the problem has existed since the WR arrived, and thinks it's because the factory had other priorities. The magnetos were formerly used for other applications, such as tractors, and were bought out and modified. So H-D had to adapt and machine the mags and their drive, and had to change and improve the wiring and the insulation. When the XRs raised the redline, the magnetos got a second bearing for the shaft, then a third.

The mags used to fire two at a time. There's a camshaft with lobes, turning at half engine speed and with the lobes placed to coincide with the staggered valve and piston sequence; but spark went to each plug when the points broke. One fired on the compression stroke and the other spark went to the other cylinder at the top of that piston's exhaust stroke, so no harm was done.

The factory hoped to improve this, and in 1978 changed to a single-fire magneto, with a rotor so the spark only went to one plug at a time, doubling the energy. Except that the rotors broke and the factory racing department offered (and sold) conversions back to the dual-fire magneto. Even then, with bearings and breakage fixed, the magnetos needed special care in isolating and insulating the plug wires, and curing the sparks that jumped (no kidding) inside and out of the mag case.

Lawwill sells ignition distributors, with one set of points for both barrels or with a set for each barrel, using car coils and a small, light, expensive battery.

Werner's system is self-powered, with tiny alternator coils, while the racing shop's system uses parts from outside, batteries and components from a California company called K/V Products. H-D doesn't sell the ignition as such. Instead it provides detailed and

Bubba Shobert, the thinking Texan. Backed by family and friends, he began winning on private Harleys. But when there wasn't enough room on that team, he joined Honda and won the national title two years in a row, 1985 and 1986, and took the record for wins in a season and wins on the mile. Warren Price/*Cycle World*

complete blueprints, instructions and part numbers so the owner can convert from magneto to electronic.

There are probably at least ten other suppliers with electronic ignitions. Some run from triggers inside the gear case, similar to the stock ignition for a Sportster, and some from triggers and rotor inside the primary case. Or, you can still buy a magneto and mount it in front of the front cylinder or on the gear case, just like old times.

In general, ignition is mostly go or no-go. If it's sour, you lose, but if it's working perfectly, all you get is the power the engine has from other places. Done carefully, magnetos work most of the time; the points-and-coil systems are easily maintained and understood, and the electronic ignitions cost more and give less trouble. Electronic is the direction people are moving but there's not much hurry, now that you almost never see or hear of an XR engine with ignition trouble.

The kit

Rotax tuner and distributor Ron Wood can't talk about the XR engine without clenching his fists. This is because he's a perfectionist.

The fastest qualifier in Grant National racing at the end of 1986, and probably the fastest combination, was this XR built by Mert Lawwill and ridden by Chris Carr. News: The huge, upside-down forks, by Simons; and, yes, the extensions below the polished portions of the sliders were bent forward. The frame tubes in front shaped closely to the engine, and the right side of the lower cradle extended outboard, so the oil pump could be removed if needed. This was the current Lawwill frame, with the oil tank part of the steering head and the backbone. That's the filler cap in front of the gas tank. The chamber in front of the front cylinder is where they cut off the old mounting for the magneto, then realized they needed the air space to accommodate crankcase pressure from the pistons' rise and fall.

Harley-Davidson isn't. The XR engine, like the KR before it, is a kit; one complete collection of the parts. When it was sold, it arrived ready to be taken apart down to the last nut and bolt, carefully checked and balanced, even magna-fluxed, before being put back together.

The winning engines get more than that. The action of the camshafts, lifting high against stiff springs, working against rockers on one side of the heads, with the heads held down by bolts through the barrels to the cases, can cause the heads and barrels to shift, move around head to barrel and barrel to case. They all must be aligned and doweled and guided and pegged into place. Same goes for the flywheels in the case halves, the halves to each other, the engine sprocket to the clutch sprocket. There are plenty of camshafts, inside the factory catalog and out, but if you want it done right you need to cut the lobes off the shafts, locate the lobes where you want them to be using precision seldom seen outside NASA, and weld them back onto the shafts. That's what Bill Werner does and he's tuned nearly fifty national race winners.

No, not every rider has a Werner on his team. Lawwill and Steve Storz, the former factory tuner now on his own in California, build engines for outside teams and riders. And they aren't the only ones. There are shops like Axtell's and Jerry Branch's, and suppliers like Carl Patrick, and even some of the larger, race-ready dealerships such as Bartel's in Culver City, California, Best's in Texas, Robison's in Daytona

Steering head for C&J frame didn't have overlap of the backbone and front downtubes, but it did have generous and carefully done gusseting. Steering damper (the thing that looks like the closer for a screen door), below the steering head and fastened to the lower clamp and frame tube, was there to control the wiggle you got on the straights when the bike worked best on the turns.

Beach, where privateers and smaller teams can get this work done right, if not cheaply.

Thing is, this work has to be done before doing any serious racing.

Lawwill's XR used his own ignition distributor, atop the gear case cover, and the cover itself and the beautifully machined alloy sprocket cover, behind the gear case. Note the expansion chamber bolted to the front of the case cover, and the breather tube in back, and that both the gearshift and rear brake pedals were on the right, one above the other. This was so the left leg was free to use as an outrigger on the turns.

An early Lawwill customer frame, with the kinked backbone braces that ran from the steering head back to the middle of the backbone. Obviously this needed its own shape of tank. Again, the trimmed-back gear case cover, but with no expansion chamber. Ignition was CD, with triggers in the housing on the gear case and power from the total-loss battery in the box at front right.

Most noticeable here is the superb workmanship: the carefully lightened and drilled engine mount plates and bracket for the footpeg, the sculptured stainless steel exhaust pipes and the nickel-plated frame. The CD ignition here is inside the plate on the primary cover. The exhaust pipes tell a story, in that the diameter and length of the pipe, and the shape of the collector and megaphone all control the engine's torque peak (where it is on the rev scale), and the width of the power band.

Engine

Most of what people consider speed parts—cams, pipes and carbs—follow what amounts to a formula: The team and top private XRs, from Werner or Peel or Lawwill, will have exhaust pipes of maybe 1.75 in. or 2.0 in. in diameter, and a length of 32 in. or 29, and they will be separate all the way to the megaphone or they'll join halfway back, according to carefully kept records. The XR, like any four-stroke, responds to certain natural laws: A short pipe boosts peak power, a longer pipe aids torque. You can vary exhaust gas velocity with the diameter of the pipe, and so forth.

Carburetors are usually Mikuni, although some Dell'Ortos and Lectrons are used. They all come in a selection of venturi sizes. Werner will use 38 mm Mikunis for the fast track where power pays off, 37s for a compromise and 36s for a TT. Jetting is as critical as you'd guess. Werner carries an air density meter so he's ready with humidity factors, and along with most of the tuners he has a complete record of which jets they used here last year, with which other settings, and how fast they went.

Gearing

The gears have become nearly an embarrassment. Back when the XR-750 came in road-race trim, gears were part of the deal, the combination and the magic.

But as the various events changed and the rules changed and H-D conceded road racing to Yamaha, and especially when the bought-out 500-R became the team vehicle of choice for TT, gearing mattered less. Much less.

So, the four-speed gearbox is pretty much as it's been since 1953, ditto the clutch and primary drive. They all work just fine, so they've remained. The catalog still has a complete selection of internal ratios for the gearbox, and there are four choices for primary drive ratio and countless variations.

The gearing changes come within the limits imposed by the number of teeth one can have on the output sprocket, and the rear sprocket, say fourteen to twenty-two for the former and thirty-five to forty-eight for the latter. The rear wheel's hub comes equipped with quick-change mounting for the rear sprocket, so most of the time the team knows weeks before the race what the best gearing for, say, the Syracuse Mile is likely to be. The primary ratio and the output sprocket will be left alone on race day, and wheels or rear sprockets will be juggled to suit. The bike itself will of course run in top gear from the middle of the first turn of the first lap, right to the flag.

Scholars may like knowing that because the Sportster has been raced so hard and well at the drags, and can be bored and stroked into displacements twice as big as the XR, the chopper side of the aftermarket is where to go for gears and shafts and gearbox parts that will absolutely stand up to any torque one can throw at them.

On the debit side, the factory had just about completed development work on five-speed internals for the XR (and by extension the Sportster) in 1973, just in time for the team to be eclipsed in road racing and not need five speeds any more.

Still another gear case cover, this time with a Werner-style double-rotor ignition where the mag drive used to be, and with a gear-driven auxiliary pump where the magneto also could have been, to drain oil from the rocker boxes and return it to the tank. This rider must have a heavy foot; the brake pedal has two pegs.

Clyde Denzer quips that the first fact concerning this gear set should be that because five speeds take the space of four, the combination "is twenty percent weaker." And presumably because ten years of unmaintained road use is harder than 200 miles of racing, the five-speed gearbox hasn't yet made it into production for the Sportster, although there are hints in that direction (and have been since 1973, too bad about that). But there have been five-speed gear sets built and used for road racing.

Oh, and the tuners talk about gearing in terms of top gear, final drive ratio only. The multiplication of engine sprocket into clutch sprocket times output into rear wheel, may be 5.34:1, so for instance, one man will say to another, "We're running five thirty-four, what are you running?"

Wheels and tires

The Class C rules have had an equalizing and stabilizing and perhaps paralyzing effect on the wheels and tires used for Grand National dirt track.

Due to historical factors already stated, the AMA began with and stayed with street-legal tires as the only choice. Then it expanded to stronger-than-street, but still with highway capability and with closely controlled tread shape and size; in other words, no knobby tires for dirt.

This has led to some oddities. The suppliers as of 1986 were Pirelli and Carlisle, with the latter taking over the role abandoned when Goodyear quit motorcycle racing, even though Continental was making approaches and submitting dirt-track tires for approval.

The 19 in. wheel is standard wear, as it has been for decades, and because that's the size used for the tires, why, the tires come in that size.

The other size, or dimension, is odd. Nearly all come labeled 4.00x19, but for example the MT-53 Pirelli used on the front will measure 3.75 in. wide and 4.0 in. high, while a Carlisle used on the back will be 4.25 in. high and nearly 4.5 in. wide. The Continental rear is so wide it won't fit all XR-750 swing arms.

So first, this has to be the only form of racing where a tire, the Pirelli MT-53, has been in the winner's circle for at least twenty-five years.

Next, the actual tire isn't as important as what the tread can be shaped into. Riders and tuners have permission to trim and shape the edges of the blocks on the tread. They use experience and intuition (and guesswork, not that they'll admit it) to work the tread into giving more grip than it would have otherwise.

Because tire size and shape is closely controlled, wheel size and shape has remained the same. There are spoked wheels and cast wheels, with rim widths of three to five inches. But while the spoked wheel in theory has some flex and the cast wheel doesn't, and a magnesium wheel is lighter than a spoked wheel with aluminum rim, neither type is clearly better than the other and it's not rare to see a bike with one of each. Racers use the mismatched wheels because the cast one had the brake they wanted or the spoked one had

Jon Cornwell's XR, with C&J single rear shock frame. You can't see the shock from here, but notice the brace triangulating the swing arm. And the twin tomahawk has been replaced by fabricated steel tubing and plate.

Another C&J, from the other side. The spring is barely visible behind and below the finned tube, which is the reservoir for the gas-charged shock inside the spring.

the rear sprocket they'll use for the heat race and it's faster to swap wheels than sprockets.

Suspension and brakes

Surely because suspension and brakes were the most recent major change in dirt track, they are where the outside influences—motocross mostly—are best seen.

The disc brake, front and rear, is universal, and all the pieces come from Grimeca or Brembo (or Honda if the budget is tight and nobody's watching).

The Girling rear shocks and 35 mm Ceriani forks have been superceded by a better selection at both ends.

Shocks now come from Fox or Simons or Works Performance, all good suppliers who began with

Tex Peel at work on Tom Maitland's bike. The rear fender here is huge and high, all the better not to create drag as the treaded tire spins inside close quarters. And the high lip might provide some downforce at speed. To the right of the tire is a catch tank from the breather, and to the right of that is the oil tank, a long, high, narrow one. Tex says it keeps the oil cooler and out of the way. The object to the left of the wheel is a special muffler, developed when noise rules were enforced; it's known as a Boom Box.

motocross equipment and thus know more about stress than the dirt crowd did. Simons and Marzocchi supply forks for most Harleys, with the new Cerianis coming into contention. All the parts are big and strong, since motocross showed the need for wheel control and torsional stiffness.

At the same time, the old idea lingers that a *real rider* rides around any shortcomings in the suspension. The Works Performance shocks have been accepted, but their designer, the man who did the single rear shock system for the Honda team, hasn't been able to do the same for Harley; the few examples of single rear shock, built by C&J, which built the Honda frames, have gone to teams that aren't the leading edge. *If* there is something to be gained from experimenting with suspension, it won't have any impact unless guys like Werner (whom O'Brien says is the best chassis man there is) or Lawwill do it successfully; then the other chaps will take notice.

Frames

This is where witchcraft comes in.

Earlier, there was mention of the quirky theory that the 45 deg. V-twin somehow had an inherent advantage over all other types of engines on dirt track because the narrow vee has some mysterious way to deliver more traction.

This has only been disproved in part; the Honda RS750 is an air-cooled 45 deg. V-twin, just like the XR. But the Honda has its connecting rods side by side instead of one inside the other, thus the crankshaft has two pins, which are offset. So, the actual piston

Sort of a team portrait. You see Scotty Parker (11) leading the tightly packed frontrunners, with Jay Springsteen (9) on the inside. This is normal riding technique for the miles: down on the paint, left hand tucked behind the number plate to reduce drag, even by a fraction, and all lined up so five or six bikes have the frontal area of one. Then, at the last possible minute, the guys in back dart out, whip around the leader and jockey for position on the turn. Harley-Davidson

sequence is 90 deg. apart, as is the firing order. The RS wins and is, at this writing, probably ahead of the XR, so the 90 deg. firing order can't be too much of a handicap, although the 90 deg. Ducati didn't do anything and the 80 deg. earlier Honda was a joke.

What's most likely here is that the secret to success in dirt-track racing is traction.

There are two rules: One, power that can be put on the ground beats power developed in the dyno room. And, two, as tuner/race team owner Carl Patrick puts it, "You want it to hook up and turn left at the same time."

That means compromise, or perhaps the best word here is balance. The weights of the engine, rider and machine have to be in the right place at the right time for cornering speed on the corners, for acceleration out of corners, and for control going into corners, which is what Patrick was saying.

Balance begins with the frame. Obviously. But that's the only obvious part because once you say the engine should be in the right place, with the correct weight distribution and wheelbase and so forth, you've got to *do* it. And no two builders or tuners or designers do it the same way, nor can any of them prove that their way is the right way.

We are working here with incremental steps. Depending on what one decides the starting point was, we go back to the KR of 1952 with the twin tomahawks for the rear mount and swing arm pivot, or to the adoption of the all-steel steering head and backbone/cradle connection and suspension with the Hiboy frame of 1967. The racers have been experimenting and working with the same outline for twenty to thirty-five years. There aren't many big questions left.

Thus, they all use a wheelbase of between 54 and 57 in. The frames all have a duplex cradle (the twin downtubes in front, below and behind the engine) and a larger, braced backbone. The material is chromoly or mild steel, and the tubes have a 1 in. diameter for

Harley-Davidson's racing team for 1986: From the left, Scotty Parker, winner of the fastest mile ever (Sacramento 1985, average speed 102.46 mph); Gene Church, reigning Battle of the Twins national champion and rider of the factory's XR1000 racer; and Jay Springsteen, three-time national champion and winner of 40 Grand National races. *Harley-Davidson*

the cradle and rear extensions, with a backbone of 1.5 or 2.0 in.

After that, they differ. The secret here is that because the general optimum has been known for years, the tiny incremental variations are critical. And they aren't that easy to see.

Compounding that problem, as power curves change and tracks change, the best combination today wasn't the best yesterday, and the settings that won the main event two years ago might put you in the second half of the fourth heat this year.

In general, Terry Knight makes the frames for Scott Parker, and a Knight frame is what you'd get if you bought one from your dealership, working out of the XR parts book.

C&J is doing the visible experimentation, with its single rear shock. C&J uses a rear mount and swing arm pivot fabricated from steel tubing instead of cast, and the engine is mounted farther forward. Don Estep, the best-placed C&J rider in 1986, says the single shock is easier to tune and that it corners better—turns in more reliably and with more control—so you can ride around the turn with a higher average speed than if you kicked the back wheel out and had to slow and then got on the power.

That may be true, but it's hard to prove. Ron Wood is a winning tuner and constructor. He had the last competitive Norton, and his Wood-Rotax singles are the best in their class. Ricky Graham is an adaptable rider, winner of the national championship on a Harley and then on a Honda. But when Graham rode for Wood, he didn't win and he never liked the Wood frame for the Harley. Wood not only didn't sell any, but the one he did build was converted into a road bike for a well-to-do sportsman.

Mert Lawwill is the most visible builder with the best kept secrets. Lawwill and Jim Belland, who later retired from motorcycle racing, did their experimenting back when frame variations were first allowed.

Lawwill's first modern frame was built in 1972, and it had a fabricated rear mount instead of the cast one. (Other tuners speculate that the twin tomahawk gets too flexible if subjected to more than 90 bhp. Lawwill chuckles and says he'd like to know how they arrived at that precise figure. But—he doesn't use the part in his frames.)

His original, modern, customer frame, made from 1977 through 1983, had two arched backbone tubes and required a different gas tank, while the other outside frames used the factory's fiberglass tank. The early frame was lighter than that of its rivals. It was short and steep, with a shorter wheelbase and pulled-in, steeper steering head angle. It worked, and Mike Kidd won the national championship with it, but it seemed harder to ride than the other frames; the current customer frame is more like what the riders expect.

The frame in the 1984 through 1987 catalog, was higher than the old one and used more trail to keep it straight. Lawwill's own racing bike, the one ridden by Chris Carr in 1986, was a full inch higher than was his 1981 racer.

Jay Springsteen used a Knight frame, the factory version modified by Werner, until 1985 when Springsteen went to a different tuner and different frame. Until then, Lawwill says, Springsteen had tried the Lawwill style but felt it moved around too much. But he won a mile with the Lawwill frame in 1985, after a three-year drought in that event.

Lawwill says the real secret is the changing tracks. They are using more chemicals, he says, and the result is a coating of what Lawwill calls "slime." This coating means less traction, which is evidenced by the gain of 15 bhp in power, while times on the mile have improved by only fractions of a second.

The fastest tracks are the ones that still have some cushion, some material that can be packed and thus geared into by the tires. The grove track, with one line and a coating of rubber, calcium, and slime, doesn't offer the grip that a cushion track does.

But either way, as traction is reduced and power increased the chassis and frame must be changed to suit. In general, higher gives more traction while lower slides better and thus is easier to ride.

The changes in frames aren't big. They aren't always noticeable or discernable because the various makes and models aren't the same when they're first

Chris Carr, who rode for the factory at one remove during 1985 and 1986. Carr's mile and half-mile machines were prepared by Mert Lawwill; his TT and short-track mounts were bought-out Rotaxes, by Ron Wood. Carr is a traditional racer, in that he came from northern California, up through the mini and club venues.

made. So, seeing a rocker box close to or far away from the front tubes, for instance, is as likely to tell you the shape, location and curvature of the engine cradle as what's been done since or how the frame's builder felt about weight distribution.

Case in point: Chris Carr was the fastest qualifier for the last two Camel Pro events in 1986. Lawwill says it came from cornering speed, not power, which could be seen by the way Carr and Honda rider Bubba Shobert walked away from the pack on the turns.

Says Lawwill, "We changed something 5/16 of an inch, and we improved by half a second a lap... It's a change that wouldn't have worked five years ago."

Just exactly what this change was, and where in the frame he made it, Lawwill naturally won't say.

Setting it up

The frame is the platform upon which and from which all other specifications and settings are made. Most of the frame designs are alike. Even the ones that weren't, though, like Ron Wood's XR with one giant spine from steering head to swing arm pivot, have used basic principles and proven principles from other venues. Aermacchi and Champion both had frames like Wood's but with different engines.

The parameters, such as wheelbase and ground clearance, are well known. The parts—for engines, suspension and brakes—are all for sale and easily found, once you know how to ask for the Lawwill or Patrick or Keen catalog (and assuming they feel like sending them out, which sometimes can be awfully slow if they don't know).

The engines can be tailored, to a point. The builders know most of what to do, and those who don't can go to tuners like Steve Storz in southern California or Lawwill in northern California and have it done: Lawwill's charge for rebuilding and preparing an XR engine, not counting head work and special

Chris Carr in action. Carr made Expert and Rookie of the Year in 1985, after winning the Junior national title, and was fourth in Grand National points in 1986. Harley-Davidson

services like magnafluxing or balancing, is $1,200, plus parts.

And there are some lucky breaks not available to outsiders: Remember the Phase Three project, the ohc engine that was never built? Some parts were made, though, and Ron Alexander got hold of the connecting rods, which were shorter than previously made from stock rods. The prototype rods found their way into Tex Peel's best miler, which Ricky Graham rode to the championship. Not gonna happen very often, or to most racers anyway.

The real secret here is that within these broad outlines and using the available parts, the best tuners, guys like Peel and Storz and Thompson and Werner and Lawwill, construct intricate combinations of the thousands of ways the XR-750 can be tuned and prepared.

You can change the wheelbase by using a longer or shorter swing arm, or by changing the engine cradle, or by using triple clamps that are closer or farther from the steering stem, or by raising or lowering the chassis on the suspension. And each change makes something else different. For instance, if you change the wheelbase you alter weight distribution and thus weight transfer under power and braking, and therefore traction.

The suspension affords more and more easily juggled settings, such as the basic damping and spring rate, and the changes that can be made at the track, with alternate mounts for the shocks and so on. If you want to go really far, copy Werner. He once took some fork sliders and filled in the axle mounts, then redrilled them off-center, so the axle and thus the wheel could be in front of the centerline or behind it, by a fraction of an inch, swapped in a second or two and made (or so he hoped) just enough change to keep the front wheel gripping.

Multiply this by hundreds, and there are countless changes to make.

State of the art

At the close of the 1986 Camel Pro year, there was an awkward sort of flurry.

First, R. J. Reynolds announced a disguised withdrawal, which meant it would sponsor some but not all of the championship events for 1987.

With that, an investigation. About the time Honda became a power in Grand National racing, people inside the AMA began wondering if there was a way to adjust the rules so Honda's obviously more powerful engine wouldn't be quite so obviously more powerful. This was done with an agreement with the guys at Honda, who don't like looking like a bully but usually managed to look like one anyway.

During the 1986 season, Kenny Roberts, the Yamaha rider who battled Harley-Davidson every chance he got when contesting the title, made a cameo appearance at the Springfield Mile . . . on Lawwill's XR. The fans from the Yamaha camp and the Harley camp loved it. At Daytona, Jay Springsteen rode a Yamaha superbike backed by laundered Daytona Speedway money, and when he rode past the Harley Owner's Group grandstand, the fans went wild, waving and cheering and whooping. In both cases the machines weren't up to the challenge, but the fans were united in their hope Honda would lose. Honda won both races.

Anyway, the AMA created a study group, borrowed some engines and did some testing at builder Jerry Branch's shop. This was an odd event, because Bill Werner sent his very best engine, and Honda said, golly, Shobert's season-winning miler had just gotten sick so here's Hank Scott's engine. Werner's XR produced 100 bhp, the best Branch had ever seen from a Harley. Scott's engine was in the low 90s, but Branch said it was tired and that he'd tested the engine earlier and it had 107 bhp; so why not agree that's what it's got?

And they did, even the press, because what all parties really wanted was to report back to the AMA that the Honda had more power (which it does, of course) so the solution should be to cut back on the power.

Shortly after that, the AMA announced the official new rules: Honda and Harley engines will use 33 mm restrictors; other makes will be restricted if they show up and win. (The official statement didn't put it that way, not in so many words, but that was clearly the meaning.) And if one make isn't slowed down, the rules will be adjusted until, by golly, they *are* slowed down.

Thus, the end of an era.

Not everybody likes this rule/restriction. OK, hardly anybody does, not out loud because if you favor changes to make the game even, why, obviously you were behind. Nobody wants to admit that.

Lawwill doesn't like the rules because the good tuners can more quickly tune around restrictors; that's what's happened with similar rules in the past. Plus, Lawwill has designed a four-valve cylinder head that in tests will let the Harley engine produce as much power as the Honda four-valve head. Lawwill's design will bolt right on. He says adding equal power makes for better, fairer racing than subtracting equal power. And that sounds right, albeit Lawwill has a lot to gain if his head design were bought by the factory and allowed to race.

Technical note: Werner's 100 bhp is probably tops with the rest of the engine as it is. But the current XR can be revved to 9200 rpm with a slim margin of safety; the limit comes from the weight and bulk of the connecting rod cages and roller bearings. If they were redone, the Harley engine could turn as fast and as long as the Honda, and with as much power.

Surprise: We have come full circle, and at the close of this fifty-year account we are back with Class A. Sure there are differences, but there are also some uncanny parallels, with the big factories fighting it out, using machines that are only partially related to what the riding public buys and rides.

Meanwhile, the racing—the classic confrontations on the fairgrounds miles, the derring-do of the open TT, the precision of short track, the head-to-head aggression of the half mile—no longer relates to what enthusiasts do. Motorcycle nuts of the eighties begin with mini dirt bikes, then go riding in the woods or desert with Dad, then try motocross or, if they're still interested when they're old enough for cars, they race motocross, or go to college and become intellectuals and then try road racing.

Grand National racing, Camel Pro racing, is in my opinion the best, closest and cleanest racing in the world, unique and distinct and special. But due to social and commercial (showbusiness) factors, it's a national pastime gone obsolete. As Samuel Goldwyn supposedly said, if people don't want to go to the movies, you can't stop them.

But the timing here is perfect. It's been a glorious fifty years. Not just for the fans, who have seen the best and the bravest in action and up close.

If you count the various changes and total up the number of all the rivals raced with and against over this time, the Harley-Davidson racing engine, the 45 deg. 750 cc V-twin, is surely the most successful and longest-lived racing engine in world history.

Despite the engine's longevity or perhaps because of it, there is no official plan to revive the racing team as it used to be. The competition department is more for supporting private riders at this writing, and the rumors of building Lawwill's four-valve heads or a really new racing engine based on the Evolution-head XL Sportster engine are . . . rumors. Race wins are down, sales are up and dirt track's best days appear to have already been.

On the human side, through all this, no matter how late the hour or empty the glasses, no matter what the circumstances or problems, not one of the guys in the factory or on the team has ever said anything against Harley-Davidson.

Coach O'Brien should be very proud of his team.

The other contenders for Best Machine and Best XR-750 look like this: Bill Werner, with Scott Parker's half miler, Ascot 1986. This is a Knight frame, wearing the latest in alloy fuel tanks (replacing fiberglass) and with rear fender/seat from Grand Prix plastics. (You can tell because the cover doesn't have visible snaps.) One of the economies of outside competition is that the seat/fender shown here retails for $125, while the factory's cataloged unit sells for $400. Werner tuned Springsteen to three national titles when the team was operating. (The company shut down the racing shop and reassigned the staff, so when the racers were outside contractors they hired their own tuners. After some dispute with management, Parker was allowed to hire Werner, on Werner's own time. Parker promptly had his best season yet, second in the series for 1986.)

Index

Ace, 12
Adamo, Jim, 165, 168
Aermacchi, 81, 89-98, 100, 102, 104, 106, 107, 108, 110, 112, 114, 116, 118, 119, 120, 127, 132-133, 138, 141, 143, 146, 150, 156, 184
Agajanian, J. C., 64, 70
Agostini, Giacomo, 145
Airhart, 85
Albrecht, Paul, 27
Alexander, Ron, 129, 133, 140, 147, 170, 185
AMA (see American Motorcyclist Association)
American Machine and Foundry (AMF), 85, 120, 153
American Motorcyclist Association (AMA), 10-12, 15-18, 26, 34, 36, 43, 46, 49, 51, 54-57, 60, 62, 68, 70-73, 75, 77, 79, 80, 81, 84, 86, 89, 94, 98, 100, 103, 109, 112, 117, 118, 120, 121, 123, 124, 126, 127, 133, 138-139, 140, 144, 149, 151, 155, 158, 160, 163, 165, 166, 169, 172, 179, 185
American Motorcyclist, 12
American Racer, 13
American Racing Motorcycles, 13, 32
AMF (see American Machine and Foundry)
Andres, Brad, 5, 45, 46, 47, 49, 54-56, 63-64, 70, 71, 83, 94, 121
Andres, Len, 5, 18, 19, 34, 44, 45, 47, 54-58, 60, 63, 69, 75, 83, 87, 121, 128, 129, 133, 135, 144
Andres, Ray, 27
Anthony, Leo, 28
Arena, Sam, 19, 32
Ascot Speedway, 64, 68, 70, 75, 80, 119, 123, 134, 136, 138, 156, 170, 171, 173, 187
Atlanta, GA, 30
Axtell, C. R., 60, 76, 82, 100, 155, 156, 177

Baja 1000, 119
Baker, Steve, 151, 156
Bartel's, 177
Bates, 116
Beauchamp, Rex, 125, 131, 155
Belland, Jim, 70, 76, 84, 85, 87, 129, 133, 137, 138, 183
Berndt, Ralph, 57-58, 60, 83
Best's, 177
Bianchi, 100
Bimoto, 109
BMW, 37
Bombardier, 118
Bonneville, 26, 105, 119, 127, 134, 136
Boody, Ted, 115, 156, 158
Boone, Mickey, 113
Branch, Jerry, 60, 76, 81, 82, 177, 185
Brashear, Everett, 54, 57
Brelsford, Mark, 5, 67, 121, 122, 129-135, 138, 141, 144, 145, 148, 149, 151, 162, 166
Breslford, Scott, 145
Brembo, 180
Brown, Bruce, 86
BSA, 37, 51, 54, 55, 56, 57, 58, 60, 68, 69, 70, 73, 74, 79, 83, 84, 89, 90, 104, 123, 128, 132, 153, 159
Buell, Erik, 166, 169
Bultaco, 71, 156
Burnett, Don, 71
Butterfield, Johnny, 30
Buzzelli, Buzz, 5

C&J, 113, 116, 174, 177, 179, 181, 183
Cactus Hare Chase, 50
Camel Pro, 117, 139, 158, 171, 184, 185, 186
Campagnale, Ben, 33
Can-Am, 117, 118
Carlisle, 179
Carr, Chris, 129, 172, 176, 183, 184
Carrillo, 62
Carruthers, Kel, 143
Castle Rock, 136
Catalina Grand Prix, 46, 51
Ceriani, 64, 73, 74, 81, 85, 86, 98, 101, 103, 104, 106, 114, 122, 127, 128, 137, 139, 140, 145, 162
Champion, 84, 116, 157, 161, 184

Chann, Pete, 30
Chevrolet, 147
Chicago, IL, 10
Church, Gene, 165, 167, 168, 182
Clark, Ken, 144, 149, 157, 162
Clymer, Floyd, 71, 149
Coates, Rod, 78
Colorado Springs, CO, 141
Columbus, OH, 50
Continental, 179
Cornwall, Jon, 179
Cosworth, 168
Cotrell, Jack, 18
Cox, Bruce, 135
Curtiss, Glenn, 9
Cycle World, 76, 87, 100, 102, 114, 148, 149
Cycle, 47, 50, 51, 54, 60, 71, 76, 81, 86, 127, 131, 132, 133, 140, 142, 143, 149
Cyclone, 10

Darr, Larry, 126, 143
Davidson, Walter, 9, 57, 58, 82, 89, 90
Davidson, William H., 46
Davidson, Willie G., 165
Davis, Jim, 56
Daytona Beach, 10, 25, 26, 27, 30, 32, 33, 34, 36, 44, 45, 47, 50, 56, 60, 65, 70, 71, 72, 73, 75, 76, 77, 78, 79, 83, 84, 85, 86, 95, 98, 99, 103, 121, 127, 128, 129, 132, 135, 136, 138, 140, 143, 145, 146, 149, 151, 153, 156, 162, 165, 166, 185
Degner, Ernst, 123
Della Santina, 5
Dell'Orto, 94, 97, 98, 101, 104, 118, 119, 178
Dennert, Bruce, 147
Denzer, Clyde, 5, 60, 72, 75, 120, 138, 140, 172, 179
DKW, 37, 89
Dodge City, KS, 10, 54
DOT, 142
Douglas Aircraft, 58
Draayer, Chris, 80
Drost, Steve, 155
Ducati, 90, 94, 165, 166, 168, 182

Duhamel, Yvon, 86
Dunlop, 107

Ecklund, Steve, 94, 158-60, 162
Edison-Splitdorf, 24
Eierstadt, Rich, 113
Elmore, Buddy, 80, 81
Elsinore Grand Prix, 46, 112
Emblem, 10
Erson, Sig, 147,
Estep, Don, 183
Evans, Jim, 143
Ezio, 99
Excelsior, 10, 12-13, 26, 69

FAM (see Federation of American Motorcyclists)
Faulk, Walt, 51, 129, 133, 135, 140
Federation of American Motorcyclists (FAM), 10
Feller, Bob, 54
Filice, Jim, 163, 164, 170
FIM (see International Motorcycling Federation)
Flanders, Earl, 60, 127, 144
Fontana, 103
Ford, 22, 30, 37
Fox, 180
France, Bill, 33
Fredericks, Curley, 11
Fulton, Walt, Jr., 75, 128

Garner, Ray, 10
Gawthrop, Rich, 151
Gibson, John, 56, 57
Girling, 84, 127, 180
Goldsmith, Paul, 50, 57, 129
Goliath, 134, 136, 141, 145
Goodyear Tire and Rubber Company, 102, 123, 142, 179
Goss, Randy, 115, 118, 160, 161, 164, 170, 171, 172, 173, 174
Gott, Randy, 120
Graham, Ricky, 94, 118, 164, 170-173, 183, 185
Grant, Ron, 86
Griffith, Jerry, 163, 164
Grimeca, 180
Guild, Cliff, 78
Gunter, Albert, 84, 123

Haaby, Don, 75
Habermill, Don, 112
Hailwood, Mike, 54, 71, 83, 132
Hammer, Dick, 87, 99, 100
Harley Owner's Group (HOG), 165, 166, 168, 169
Harley, Arthur, 9
Harley, William J., 9, 55
Harley, William S., 9, 55, 79
Hatfield, Jerry, 5, 32, 93
Hepburn, Ralph, 26
Hercules, 26
Hill, Bobby, 30
Hocking, Rick, 149
Hodaka, 98
Hoecker, Bill, 5
Hoffer, Eric, 77
HOG (see Harley Owner's Group)
Hollingsworth, Don, 98
Hollister, CA, 18

Honda, 45, 78, 90, 93, 97, 98, 112, 116, 117, 119, 127, 128, 144, 145, 149, 152, 153, 162, 164, 170-174, 181, 183, 184, 185
Houston, 114, 115, 116, 143, 145, 156, 164, 173
Huber, Billy, 30, 45, 86
Husqvarna, 112

Indian, 10-13, 15, 19, 23, 30, 32, 33, 34, 55, 58, 69, 144, 153, 163
Indianapolis, 81, 151, 154, 161, 163
Ingraham, John, 112
International Motorcycling Federation (FIM), 71, 145
Isle of Man, 71
Itoh, Fumio, 50

Jackpine Enduro, 89
Jarrell, Randall, 149
Jennings, Gordon, 81, 84, 87
Jones, L. J., 5
Jorgensen, Alex, 94, 118, 119

Kathcart, Griff, 16
Kawasaki, 86, 99, 104, 143, 149, 152, 153, 165, 171
Kayaba, 113
Keen, 184
Keener, Corky, 151, 155, 158, 163
Kidd, Mike, 143, 163, 164
Klingsmith, Jack, 151
Knievel, Evel, 129
Knight, 115, 116, 119, 183
Kretz, Ed, 19, 32
Kroll, Matt, 138
KTM, 118, 119
Kuchler, Lin, 79
Kuldowski, Don, 113
K-V Products, 175

Laconia, NH, 30, 34, 49, 54, 56, 70, 71, 81, 99
Laguna, Seca, 109
Langhorne, PA, 47, 54, 56
Lawson, Eddie, 144
Lawwill, Mert, 5, 38, 60, 68, 71, 74, 76, 79, 80, 81, 82, 83, 84, 85, 86, 99, 122, 124, 127, 129, 131, 132, 133, 134, 136, 137, 138, 141, 143, 145, 147, 148, 150, 159, 164, 172, 175, 176, 177, 178, 181, 183, 184, 185, 186
Lectron, 178
Leonard, Joe, 45, 46, 50, 54, 57, 58, 64, 67, 71, 93, 94, 98, 133
Lewis, Paul, 168
Lincoln, IL, 59
Linkert, 53
Lipp, 116
Livonia, MI, 136
Long Island Endurance Run, 9
Long, Jerry, 151
Louden, NH, 132
Louisville, KY, 141
Lucifer II, 166
Lucifer's Hammer, 162, 166-169
Lyons, Pete, 159

M&ATA (see Motorcycle and Allied Trades Association)
M.G., 34, 36
Maiers, Larry, 174

Maitland, Tom, 180
Mamola, Randy, 83, 144
Mangham, James "Stormy," 105
Mann, Dick "Bugsy," 38, 58, 69, 70, 71, 74, 75, 77, 84, 98, 123-124, 128, 131, 133
Manning, 136
Mantle, Mickey, 54
Markel, Bart, 54, 55, 59, 60, 63, 64, 68, 69, 70, 72, 75, 76, 77, 79, 80, 83, 84, 93, 98, 114, 122, 131, 132, 133, 143, 160
Marzocchi, 118, 162, 167, 181
Matchless, 50, 71, 74, 78, 79, 84
McKay, Jerry, 18
McPhee, John, 117
McQueen, Steve, 136
Mikuni, 101, 119, 138, 141, 178
Milburn, Bill, 5, 69, 75
Monza, 107
Moss, Stirling, 54
Moto Guzzi, 166
Moto-X Fox, 167
Motorcycle Weekly, 135
Motorcycles and Allied Trades Association (M&ATA), 10
Motor Maids, 149
Muzzy, Rob, 171

NASCAR (see National Association for Stock Car Automobile Racing)
National Association for Stock Car Automobile Racing (NASCAR), 168
Nazareth, PA, 86
Neal, Don, 47
Neilson, Cook, 127, 130, 143
Nelson Ledges, OH, 98
Netz, Louis, 165
Nichels Engineering, 129
Nix, Fred, 75, 80, 81, 82, 83, 84, 85, 86, 87, 122
Nixon, Gary, 70, 81, 82, 84, 87, 131, 143, 148-149
Norton, 33, 34, 37, 60, 64, 71, 106, 118, 130, 144, 159, 166, 183
NSU, 89

O'Brien, Dick, 5, 45, 57, 58, 59, 60, 68, 74, 76, 80, 81, 82, 83, 84, 85, 86, 89, 99, 107, 110, 120, 121, 123, 124, 127, 129, 130, 133, 134, 136, 143, 146, 150, 151, 153, 154, 156, 166, 167, 168, 169, 170, 172, 181, 186
Offenhauser, 126
Ogilvie, Bruce, 119
Oldani, 71, 72, 78, 80, 95
Oldsmobile, 38
On Any Sunday, 86
Ontario, 147, 150

Packard, 37
Palmgren, Chuck, 38
Palos Verde, 49
Panther, 116
Parilla, 99
Paris-Dakar Rally, 112
Parker, Scott, 115, 124, 172, 181, 182, 183, 187
Pasolini, Renzo, 132, 143, 144, 148, 150
Patrick, Carl, 177, 182, 184
PB, 67
Peel, Tex, 5, 118, 164, 170, 171, 173, 178, 180, 185
Peoria, 10, 51, 98

Perkins, Dud, 18, 70, 71, 84, 129
Petrali, Joe, 20, 26, 30
Pickrell, Ray, 135
Pink, Don, 90
Pirelli, 179
Plymouth, 37
Pohlman Studios, 28
Pontiac, MI, 172
Pope, 10
Pratt, Ted, 106

R. J. Reynolds Tobacco Company, 55, 157, 185
Rainey, Wayne, 83
Rall, Ronnie, 98
Rancourt, Norris, 99
Rayborn, Calvin, 38, 72, 73, 75, 76, 80, 81, 82, 83, 84, 85, 86, 95, 99, 121, 122, 128, 129, 131, 135, 136, 140, 141, 143, 144, 148, 166
Read, Phil, 83
Reiman, Roger, 61, 68, 70, 71, 75, 77, 78, 79, 82, 83, 85, 87, 99, 105, 128, 133
Resweber, Carroll, 5, 7, 50, 52, 54, 57-60, 64, 67, 69, 71, 82, 86, 93, 98, 123, 129, 133, 162, 166, 172
Reynolds (see R. J. Reynolds Tobacco Company)
Ridder, Ken, 70
Riley, Warner, 136
Riviera, 136
Road Atlanta, 132, 140
Roberts, Kenny, 83, 102, 103, 104, 143, 144, 145, 146, 149, 151, 153, 156, 157, 158, 162, 164, 185
Robinson, Dot, 149
Robinson, Earl, 149
Robison's, 177
Rockwood, Roxy, 73, 78, 84
Roeder, George, 58, 63, 69, 71, 75, 80, 81, 82, 103, 105, 135
Romero, Gene, 86, 94, 128, 129, 171
Rotax, 116, 118, 119, 172, 176, 183
Rouitt, Dan, 95

Saarinen, Jarno, 143, 144
Sacramento, CA, 182

Salzburgring, 107
San Jose, CA, 19, 30, 148, 150, 151, 172
San Mateo, CA, 51
Santa Fe, NM, 157, 161
Savannah, GA, 33
Schwinn, 11
Scott, Gary, 102, 103, 104, 107, 109, 110, 143, 145, 151, 154, 155, 156, 164, 170, 185
Scott, Hank, 160
Sehl, Dave, 131, 141, 143, 147, 148, 149
Shell, 157
Shobert, Bubba, 51, 170, 171, 172, 175, 184
Sifton, Tom, 19, 30-32, 34, 45, 46, 51, 54, 56, 57, 60
Simons, 176, 180, 181
Skoal Bandit, 168
Smart, Paul, 131
Smith, E. C., 15, 56, 79, 149
Smithsonian Institution, 26
Smitty, 126
Sonicweld, 84
Spear, Pat, 18
Spencer, Freddy, 83, 144, 163, 164
Springfield, IL, 10, 34, 45, 47, 60, 185
Springsteen, Jay, 94, 103, 114, 115, 124, 129, 151, 154-65, 167, 169, 172, 181, 182, 183, 185
Staten, Rex, 113
Storz, Steve, 5, 177, 184, 185
Sturgis, SD, 51
Surtees, John, 64
Suzuki, 83, 86, 104, 110, 123, 144, 145, 165
SWM, 118
Syracuse, NY, 151, 178
Syvertson, Hank, 20, 45, 57, 58

Tancrede, Babe, 33
The Motorcyclist, 15
Thompson, Brent, 5, 103, 112, 161, 185
Thuett, Shell, 145, 156
Tiger, 78
Tilley, Don, 167, 169
Tillotson, 53, 57, 83, 127
Torres, Ben, 18

Torrey Pines, 54
Trackmaster, 95, 114, 116
Tripes, Marty, 113
Trippe, Gavin, 135
Triumph, 34, 37, 55, 56, 57, 60, 68, 71, 73, 78, 79, 80, 81, 82, 83, 84, 86, 89, 94, 104, 128, 129, 131, 135, 144, 148, 153, 154, 156, 159, 166, 171
Tuttle, Mark, Sr., 112

United States Motorcycle Club (USMC), 26, 71
USMC (see United States Motorcycle Club)

Velodrome (see San Jose, CA)
Villa, Walter, 100, 102, 107, 108, 110
Vincent, 67, 153
Vista-Sheen, 151, 164

Watkins Glen, 64
Weatherly, Joe, 30
Werner, Bill, 5, 131, 133, 147, 150, 151, 154, 156, 160, 161, 163, 164, 174, 178, 181, 185, 187
White, Ralph, 69, 75
Wico, 21
Widman, Earl, 127
Willow Springs, 54, 70
Windber, PA, 51
Wixom Brothers, 73, 82, 116, 127
Wood, Ron, 117, 118, 119, 172, 176, 183
Work, Bob, 151
Works Performance, 180, 181
Wrecking Crew, 10
Wright, Steve, 5, 87

Yamaha, 45, 79, 83, 86, 98, 100, 104, 106, 112, 113, 117, 143, 144, 145, 146, 149, 151, 153, 156, 157, 159, 162, 163, 164, 165, 170, 185

Zundapp, 90
Zylstra, Pieter, 5, 124, 130, 138, 153, 175